AF560289

Post-harvest Handling of Food Crops

Post-harvest Handling of Food Crops

Vikram Sharma

Post-harvest Handling of Food Crops

ISBN 978-93-5111-522-9

Published in 2015 in India by

RANDOM PUBLICATIONS

4376-A/4B, Gali Murari Lal, Ansari Road
New Delhi-110 002
Phone : +9111-43580356, 011-23289044, 011-43142548
e-mail: sales@randompublications.com,
info@randompublications.com, randomexports@gmail.com

Reprint 2021

Type Setting by : Friends Media, Delhi-110089
Printed at: Replika Press Pvt. Ltd.

Preface

In agriculture, post-harvest handling is the stage of crop production immediately following harvest, including cooling, cleaning, sorting and packing. The instant a crop is removed from the ground, or separated from its parent plant, it begins to deteriorate. Post-harvest treatment largely determines final quality, whether a crop is sold for fresh consumption, or used as an ingredient in a processed food product. Effective handling decreases post-harvest losses.

The most important goals of post-harvest handling are keeping the product cool, to avoid moisture loss and slow down undesirable chemical changes, and avoiding physical damage such as bruising, to delay spoilage. Sanitation is also an important factor, to reduce the possibility of pathogens that could be carried by fresh produce, for example, as residue from contaminated washing water.

After the field, post-harvest processing is usually continued in a packing house. This can be a simple shed, providing shade and running water, or a large-scale, sophisticated, mechanized facility, with conveyor belts, automated sorting and packing stations, walk-in coolers and the like. In mechanized harvesting, processing may also begin as part of the actual harvest process, with initial cleaning and sorting performed by the harvesting machinery.

This book deals with the post-harvest handling of food crops to reduce or prevent the post-harvest losses. It will be an invaluable resource for research professionals, quality control personnel, and post-harvest biology students anyone involved in the technology for handling and storing fresh fruits, vegetables, and ornamentals.

Author

Contents

1

Post-harvest Losses

Over the past quarter of a century, since the appearance in the 1960s of what has been called the Green Revolution, many writers and agri-food specialists throughout the world have addressed the problem of the losses occurring at various stages in the food chain or pipeline. The first World Food Conference, held in Rome in November 1974, when the African countries of the Sahel had started to suffer from drought and famine, proved an opportunity to draw attention to the gravity of the issue.

From time to time, people feel called upon to criticize the lack of solid results in the sphere of post-harvest losses. While these losses are undoubtedly of major importance, issue must be taken with the style of such criticisms, which are too often based on incomplete or poorly understood data, or mean that certain pre-harvest shortcomings are conveniently ignored.

In view of the questions asked and the articles and books dealing with the subject, it therefore seemed vital to challenge accepted ideas and highlight the complexity of the issues involved in post-harvest losses.

First of all, people too often forget that loss, in the sense of decline and decay, is an integral part of the life cycle of both individuals and societies. At a time when humanity is becoming more aware of its collective responsibility toward the environment, it would be paradoxical if it forgot that food is a living thing, which evolves within a complex of factors, not only biological and climatic, but also human, i.e. economic, social, cultural and political.

This does not mean that differences in climate should be ignored, for they constitute major constraints over which we have little control. Unlike developed countries which are mostly situated in temperate zones, developing countries are in tropical zones and are exposed either to one short rainy season, or to frequent and sometimes violent rainfall and high humidity. Such conditions have major consequences for post-harvest systems, especially the storage stage. The combination of heat (around 30°C) and ambient humidity (between 70 and 90 percent) fosters reproduction of the main insect pests, which multiply fast (up to 50 times in a month for the bean-seed beetle and 70 times for the Tribolium or flour beetle), as well as micro-organisms such as yeasts and moulds. Respiration is also speeded up, so that nutrient reserves suffer and premature germination can even begin, thereby compromising the preservation and future use of produce.

Weather conditions alone cannot explain the infestation and proliferation of parasites and pests in foodstuffs. International trade also plays a part. People often cite the case of rats which travel aboard cargo-ships and then escape when the ship docks finding their way into warehouses. However, insects also migrate from one continent to another and some of these "hidden" migrations have had disastrous consequences, such as the recent case of the larger grain borer (Prostephanus truncatus), which is discussed later. There is a danger that the enormous increase in intercontinental travel and trade of every kind can only increase this kind of risk.

The effects of new seed varieties developed in research centres and laboratories should also be taken into account. They not only require more investments and various products (seeds, irrigation, fertilization, phyto-sanitary treatments) involved in cultivation and production, but are also more vulnerable to the depredations of pests and to damage in post-harvest handling. Agricultural research, with the help of genetics, has given priority to plant productivity, but this quantitative concern has entailed new difficulties in connection with harvesting (bent or flattened stalks, tough stalks, need for artificial drying, etc.), threshing (breakages, grain fragility) and especially preservation (greater sensitivity to atmospheric fluctuations, pest attacks, etc.). Successful adaptations are not produced overnight and are never made without the support and cooperation of growers.

Sustainability through repeated cycles of germination and ripening is vital, although there are contradictions here too. At a time when sustainable farming practices are the centre of attention, the world of economics is dominated by a concern for short-term and preferably, maximum yields. This does not sit well with farming, which does not generate high short-term profits but is oriented more toward long-term results.

Those who are disturbed by the level of losses often base their reasoning on economic calculations showing the profits possible if these losses were eliminated, without taking the trouble to examine the costs of arriving at such a result. Taking harvesting as an example, a combine-harvester can easily be replaced by an army of harvesters who would gather each ear by hand, but at what cost? Would such an effort be considered free and for whom? Peasant farmers' lives are far from easy and it is hard to imagine that they are squandering their harvests unnecessarily. Another point often overlooked is the effects of losses on production. A 20 percent harvest loss means that 20 percent of the elements necessary for production have also been lost, whether labour, inputs, water or time. This point is very seldom mentioned in estimates of losses, at a time when maximum efficiency is sought and available arable land and water are shrinking.

Taken together with the lack of certainty on the source of figures, such oversights and omissions are puzzling, but they do in fact reflect the more general problem of the reliability of available information. The quest for greater rigour and consistency raises a methodological problem. In the first place, the basis for any analysis has to be defined. What operations are included in the post-harvest system? What is its starting point and where does it end? On what basis are losses to be calculated, amounts harvested, hoped for, or estimated? Do we work from the theoretical or recorded yield per hectare? Are post-harvest operations part of the production cycle, or should they be considered part of the processing and distribution sector?

All too often, activities tend to stop when the product is ripe, with a marketing or processing component tacked on as an after-thought, with no concern over continuity and consistency in approach. What happens between the field and the trader's shop or the processing plant? At what stage are rights of ownership transferred from one operator to the next? How should the added value be apportioned?

An overall understanding of the workings of the post-harvest system is clearly needed and a number of questions still requires answers. It therefore seems best to review the problem as a whole and work out a methodology that can lead to solutions. This approach would have a different aim from the majority of surveys and studies carried out to date. Rather than focusing on the observation and calculation of losses in each operation, it would give priority to identifying the factors responsible for these losses through all the phases of the post-harvest chain. It would be descriptive, analytic and explanatory, rather than quantitative and hence, more in keeping with both the spirit and understanding of a complex whole in which technological, sociological and economic elements are closely interwoven. This would provide a more credible and solid basis for developing possible technical improvements (human position and intervention, equipment, structures, materials, etc.) and introducing effective innovations.

While awaiting these changes, small producers will continue to protect their harvests and preserve their produce as best they can, slowly and cautiously adopting new methods, as they have done throughout the centuries, carefully refining techniques and systems to meet their own specific needs. Traditional granaries are a good example: genuine models of local architecture and functional craftsmanship, their effectiveness has been questioned in recent years. Developers have generally tended to take this latter view, believing they were doing the right thing in proposing only systems based on those of developed countries, without trying to prepare the ground for progressive incorporation of new techniques. They were certainly unaware of what Martin Greeley, a man with an excellent understanding of Asian cultures and peoples, was writing at the time: "It is simply unreasonable that farmers who have developed their post-harvest system over centuries should be suffering food losses of 20 percent or more". Such an observation should be borne in mind when planning and carrying out any action aimed at developing the rural and agricultural sector.

Food supplies for both animals and human beings are being put at risk as people blindly ignore facts and pursue a competitive, productivity-centred line of economic reasoning. Moreover, while consumption is rapidly growing, in many countries more and more farmers are abandoning the land. There are clear grounds for admitting that the old peasant farmers had and still have, a finely honed sense of economics when they fed their food waste and produce, unfit for their own consumption, to their livestock. This was their way of building up capital on the hoof, as well as ensuring themselves a liquid reserve and a prized food for special occasions. Not only were post-harvest losses limited, but anything that could not be eaten was put to some other use, thus providing some guarantee against emergencies and unforeseen events.

Definition of the Post-harvest System

"Properly considered," Spurgeon says, "the post-harvest system should be thought of as encompassing the delivery of a crop from the time and place of harvest to the time and place of consumption, with minimum loss, maximum efficiency and maximum return for all involved."

The term "system" denotes a dynamic, complex aggregate of logically interconnected functions or operations within a particular sphere of activity. The term "chain" or "pipeline" highlights the functional succession of various operations but tends to ignore their complex interaction.

In considering the system or the agro-food chain as a whole, harvesting can be seen as the hinge, or as a ridge between the pre-harvest slope, corresponding to production activity and the post-harvest slope, extending from harvesting to consumption.

The post-harvest system encompasses a sequence of activities and operations that can be divided into two groups:

— technical activities: harvesting, field drying, threshing, cleaning, additional drying, storage, processing;

— economic activities: transporting, marketing, quality control, nutrition, extension, information and communication, administration and management.

Elements of the Post-harvest System

Harvesting

The time of harvesting is determined by the degree of maturity. With cereals and pulses, a distinction should be made between maturity of stalks (straw), ears or seedpods and seeds, for all that affects successive operations, particularly storage and preservation. Pre-harvest drying, mainly for cereals and pulses. Extended pre-harvest field drying ensures good preservation but also heightens the risk of loss due to attack (birds, rodents, insects) and moulds encouraged by weather conditions, not to mention theft. On the other hand, harvesting before maturity entails the risk of loss through moulds and the decay of some of the seeds.

Transport

Much care is needed in transporting a really mature harvest, in order to prevent detached grain from falling on the road before reaching the storage or threshing place. Collection and initial transport of the harvest thus depend on the place and conditions where it is to be stored, especially with a view to threshing.

Post-harvest Drying

The length of time needed for full drying of ears and grains depends considerably on weather and atmospheric conditions. In structures for lengthy drying such as cribs, or even unroofed threshing floors or terraces, the harvest is exposed to wandering livestock and the depredations of birds, rodents or small ruminants. Apart from the actual wastage, the droppings left by these marauders often result in higher losses than what they actually eat. On the other hand, if grain is not dry enough, it is vulnerable to mould and can rot during storage.

Moreover, if grain is too dry it becomes brittle and can crack after threshing, during hulling or milling. This applies especially to rice if milling takes place a long time (two to three months) after the grain has matured, when it can cause heavy losses. During winnowing, broken grain can be removed with the husks and is also more susceptible to certain insects (e.g. flour beetles and weevils). Lastly, if grain is too dry, this means a loss of weight and hence a loss of money at the time of sale.

Threshing

If a harvest is threshed before it is dry enough, this operation will most probably be incomplete. Furthermore, if grain is threshed when it is too damp and then immediately heaped up or stored (in a granary or bags), it will be much more susceptible to attack from micro-organisms, thus limiting its preservation.

Storage

Facilities, hygiene and monitoring must all be adequate for effective, long-term storage. In closed structures (granaries, warehouses, hermetic bins), control of cleanliness, temperature and humidity is particularly important. Damage caused by pests (insects, rodents) and moulds can lead to deterioration of facilities (e.g. mites in wooden posts) and result in losses in quality and food value as well as quantity.

Processing

Excessive hulling or threshing can also result in grain losses, particularly in the case of rice (hulling) which can suffer cracks and lesions. The grain is then not only worth less, but also becomes vulnerable to insects such as the rice moth (Corcyra cephalonica).

Marketing

Marketing is the final and decisive element in the post-harvest system, although it can occur at various points in the agro-food chain, particularly at some stage in processing. Moreover, it cannot be separated from transport, which is an essential link in the system.

Table 1. Comparison between properties of cereals and roots and tubers regarding their storage capacity

Non-perishable food crops	*Perishable food crops*
Harvest manly seasonal, need for storage of long duration	Possibility of permanent or semi-permanent production, needs for short-term storage
Preliminary treatment (except threshing) of the crop before storage exceptional	Processing in dried products as an alternative of the shortage of fresh products
Products with low level of moisture content (10-15 percent or even less)	Products with high level of moisture in general between 50-80 percent
Small "fruits" of less than 1 g	Voluminous and heavy fruits from 5 g to 5 kg or even more
Respiratory activity very low of the stored product, heat limited	High or even very high respiratory activity of stored products inducing a heat emission in particular in tropical climates
Hard tissues, good protection against injuries	Soft tissues, highly vulnerable
Good natural disposition for storage even for several years	Products easily perishable, natural disposition for storage between some weeks up to several months (strong influence of the varieties)
Losses during storage mainly due to exogenous factors (moisture, insects or rodents)	Losses due partly to endogenous factors (respiration, transpiration, germination) and partly to exogenous factors (rot, insects)

Post-harvest Losses

Losses are a measurable reduction in foodstuffs and may affect either quantity or quality. They arise from the fact that freshly harvested agricultural produce is a living thing that breathes and undergoes changes during post-harvest handling.

Loss should not be confused with damage, which is the visible sign of deterioration, for example, chewed grain and can only be partial. Damage restricts the use of a product, whereas loss makes its use impossible.

Some basic definitions are needed before moving on to the various types of loss.

Foodstuff

Products, in the present case crops, edible by human beings; more specifically, the part fit for human consumption. In tropical countries, 75 percent of basic food comes from cereals and pulses. The remaining vegetable-based food is often, especially in wet, wooded zones, supplied by roots and tubers, particularly cassava, yam, taro, plantain, potato and sweet potato. In the food chain, quantities of food are usually expressed in terms of weight but this does not mean that organic structure and nutrients can be ignored.

Grains and Seeds

Cereals, pulses and oilseeds grown in most climates and latitudes for human consumption. The main cereals are wheat, maize, rice, barley, sorghum, millet, oats and rye; pulses cover the various species of pea, bean, broad bean and lentil; and oilseeds cover soya, groundnut, sesame, rapeseed and sunflower.

Post-harvest

If harvesting covers the period when the various products grown are removed from the field, after maturity, the post-harvest period runs from exit from the field to the time of culinary preparation. For various reasons, but especially to allow the straw and grain to dry fully, harvesting may be delayed sometimes for months, as happens particularly with maize and rice and in these cases, some people prefer to speak of "post-production" in order to indicate the link between harvesting and post-harvest operations.

Food Loss

Food loss refers to total modification or decrease of food quantity or quality which makes it unfit for human consumption.

Types of Losses

The first distinction in agro-food losses is that between quantity and quality. Quantitative loss is a loss in terms of physical substance, meaning a reduction in weight and volume and can be assessed and measured. Qualitative loss, however, is concerned particularly with the food and reproductive value of products and requires a different kind of evaluation.

It should be noted that we shall not consider losses occurring during the production period and caused by various crop pests (insects, weeds, disease), even though they have a major influence on food preservation conditions and account in part for the nature and size of post-harvest losses.

After describing direct and indirect losses, we shall thus look at weight loss, then quality loss, food loss, seed viability loss and commercial loss. We should not forget the existence of moisture content and the difference between damage and loss.

Moisture Content

In bio-chemical terms, organic products are composed of dry matter and water. The moisture content is the amount of free water within a given product and is expressed either as a decimal proportion or a percentage. For example, with cereals, a 13 percent moisture content is considered a guarantee of satisfactory grain preservation. In agriculture, moisture content or humidity rate is usually indicated as a proportion of the moist product; i.e. the moisture content is the proportion of the weight of moisture to the total weight of dry matter and moisture.

Damage

Damage is a clear deterioration in the product, e.g. broken or pitted grain, which affects more its quality than its quantity and can in the long-term result in a definite loss. Both damage and loss should be quantified in terms of weight and cost.

Direct and Indirect Losses

Direct losses occur when the disappearance of a foodstuff is caused by leakage (for example, spillage from bags) or consumption by pests (insects, rodents, birds), whereas indirect losses occur when a reduction in quality leads to the consumer's refusal to purchase.

Weight Loss

While weight loss is easy to observe and measure, it does not necessarily mean food loss, since it can result simply from a reduction in moisture content. Moisture loss during drying is therefore not a food loss. On the other hand, an abnormal increase in weight through moisture absorption following rainfall on stocks left in the open can cause serious damage resulting in loss.

Weight loss can be caused by leakage, during transport for example, if sacks have holes or are insecurely attached. It is often the result of prolonged infestation and consumption by insects, rodents and birds or poor packaging. Weight loss from pests is not immediately apparent and may deceive an inexperienced purchaser. It can be checked by taking an equivalent amount of clean, healthy cereal, milling the two samples and weighing the flour from each. The poorer sample will produce less flour. This method can also be used to check whether the weight is really correct, for it is easy to increase it by moistening the grain or adding foreign bodies such as pebbles, earth or waste material.

Quality Loss

Quality criteria cover a wide range and are concerned both with external features, shape and size and with odour and taste. The cultural factors that can influence diets and food habits must also be borne in mind.

The cleanliness and healthy condition of a product are primary concerns for the market and correspond to what is referred to as a "sound, legal and merchantable" product in commercial law. For example, if a trader takes a handful of grain from a sack, he can quickly see if it releases dust and work out if this is the result of insect infestation. Similarly, a bad smell can arouse suspicions that rodents have been at the grain, verifiable by the presence of rat or mouse droppings and hair. Many other objects can be mixed in with a foodstuff and reduce its value: bad grain, scraps of straw or other plant residue, soil, pebbles, bits of glass, etc. All such objects are hard to remove but some of them represent a greater risk of contamination than others: the soluble excreta of pests, oils, pesticides, pathogenic organisms spread by rodents and toxins from fungi and moulds.

It goes without saying that the presence of foreign bodies, which can distort the weight of a batch being sold, also affects the quality and thus the market value of a product.

Food Loss

While food loss clearly results from a loss in quantity, it also results, but more insidiously, from a loss in quality and edibility, making it unfit for human consumption. Staple foods contain not only essential nutrients but also important vitamins. For instance, grain hearts are rich in protein and vitamins; however, they are also the preferred target of rodents and insects.

The various nutritive parts of products are thus the prey of different families of parasites. Weevils feed especially on the endosperm, the inside of the seed, which is rich in carbohydrates, while many parasites attack the cereal cover, which is rich in vitamins. Vitamin content is also affected by humidity during storage and by mould infection.

Seed Viability Loss

Seed set aside for sowing, like any product used for reproduction, is preserved with great care in order to maintain its full germinative potential. As noted above, the protein-rich grain heart can be a favourite target of certain pests. Atmospheric conditions also play a part, as they can weaken the seed's productive potential; variations in light, temperature and humidity, leading to excessive respiration, are particularly responsible here.

Commercial Loss

Commercial loss is the translation of the various types of loss listed above into economic and monetary terms. Although the price of a foodstuff is usually based on weight, many other factors play a part. This applies especially to the qualitative elements emphasized above, starting with cleanliness and purity, which will be all the more sought after if supplies are abundant in the market-place.

Reference to abundance brings us to the key economic factor of the supply-and-demand situation at a given moment.

A producer's ability to take advantage of a time when the scarcity of a commodity pushes up the price, is an element in good marketing management of his production, based on information and foresight. This assumes good organization, in which structural and not only market-situational factors are taken into account (for example, having enough storage facilities available and in good condition to be able to hold on to the harvest until the price rises) as well as keeping abreast of long-term developments, such as socio-cultural changes, which affect the future.

Although this kind of market-oriented attitude depends on an individual's skills, it also has some collective aspects that deserve development. For the present, we would note that the ability to capitalize on a product depends on a number of technical and economic factors, in which the quality of the merchandise plays an increasingly important role but so also do the human qualities of the producer and hence, the training that readies him to take his place as an economic player in the life of his country.

Irreducible Losses and Compensation

While loss of weight during drying is normal and measurable, there are other, said "irreducible" losses, which arise basically from respiration of the product and mechanical rubbing of grain against itself, as well as the breakage inevitable with certain machines. It should therefore always be remembered that losses, whether pre-harvest or post-harvest (i.e. in production, distribution, storage or marketing) cannot be materially reduced to zero and that they have to be compensated for through extra production. The production increase rate must be progressively higher than that of losses, if such compensation is to be adequate, so that compensation for a 20 percent loss will require 25 percent more production, for a 40 percent loss, 66 percent more and for a 60 percent loss, 150 percent more.

Study of Post-harvest Losses

Study of post-harvest losses covers many aspects, in view of the wide range of (a) products involved; (b) successive operations in the post-harvest system; (c) causes of losses; and (d) pests and other food parasites, not to mention; (e) physical, technical, economic and other conditions that aid and abet the action of the agents of deterioration, consequently increasing losses. This has led to the enormous variety in approach and form of analysis found in reports, manuals and other works, depending on the authors' choices and aims. For example, M. Bourne focuses on the causes of loss, distinguishing primary and secondary causes and the places and times of loss; thus he examines biological and microbiological causes first and is primarily interested in insects, rodents and moulds. Similarly, a recent NRI training manual, which does not distinguish between primary and secondary causes, starts by considering pests and micro-organisms and then moves on to losses during successive post-harvest operations.

G. Schulten takes a more selective approach in a FAO mission report, further differentiating between several categories of produce. Beginning with rice, which he distinguishes from other cereals, he argues that losses for this crop are a result primarily of handling and equipment, particularly at the time of harvesting. The author therefore takes a close look at all the manual and mechanical operations from on-field drying through to paddy milling. Only then does he move on to a comprehensive consideration of cereals and pulses in order to analyse loss during storage, distinguishing small-farmer or village storage from warehouse storage and confining himself to loss caused by insects and moulds. Rodent damage is studied separately.

In the above-mentioned NRI manual, losses are examined in sequence, following the succession of operations in a single system, with special attention to the conditions of cutting and field drying of rice ears and then to the risks of damage and loss during milling.

Most recent reports and handbooks pay due attention to the question of equipment in general and more specifically to small-scale mechanization, particularly as used in threshing and hulling processes. They also take account of some of the financial, technical and economic effects these may have. However, very few pay any significant attention to socio-economic factors or socio-cultural aspects (traditional division of labour, dietary habits, cooking methods, etc.).

This brief outline of the non-technical aspects of post-harvest losses reminds us that the post-harvest sector itself is part of a much broader and more complex system, in which the various types of interrelationship and interdependence concern not only the successive operations in the food chain, but also the whole range of human activities and hence the very workings of society, thereby contributing to existence and the advance of civilization.

Loss Assessment and Field Surveys

Rice

Many of the available works in loss assessment, based on field surveys, are devoted to rice-growing and its post-harvest system. D. Calverley's regional assessment, covering several countries in central and south-eastern Asia, is a good example and gives an idea of the complexity of a summary and/or comparative evaluation, whether one or more products are being considered. Since rice holds pride of place, we shall start with this cereal, so that comparison will focus mainly on working methods, implements and geography. Furthermore, we shall confine ourselves for now to results connected with harvest operations as such: maturity at the time of harvesting, manual or mechanical cutting (or gathering), placing in stacks and bundles, on-field and off-field drying of paddy.

Given the diversity of the causative factors in this example, it would be hard to draw any general conclusion valid for all the countries concerned.

Doubt may be voiced on the validity of over-simplified results and all-embracing statistics such as these, primarily on the basis of the lack of consistency in the choice and presentation of the technical operations or headings and sub-headings considered. Such a disparity is clearly a result of differences in the research and survey methods adopted, quality of field teams, means at their disposal, priorities set for their work and goals pursued. However, it is also a result of differences in levels of social and economic development of the rural and agricultural sector, particularly in terms of improved techniques and the introduction of modern equipment (machines, motors), not to mention the use of new varieties or selected seed and chemical products for fertilizing crops and treating harvests.

A three-year study conducted in China divides harvesting into two categories (sickle reaping and combine-harvesting), threshing into another two (pedal and motor threshing), drying and cleaning into three sub-groups (sun-drying on bamboo, on cement and with a screen) and storage into three causes of loss (moulds, insects and rats). With ten sub-headings, this provides a more detailed and more specific table in which the loss in each sub-operation further explains the average loss for each major operation.

Given the high percentage of moisture in the grain and the damp heat of the tropical forest, there can be considerable damage and loss if pre-harvest field drying is extended, especially starting in the third month, when the depredations of insects and birds, added to the percentage of fallen stalks (broken or flattened), can amount to 50 percent of the harvest, while the moisture content of the grain is still too high for long-term storage (the maximum levels are 14 percent for paddy and 13 percent for milled rice).

We can close these observations on rice by noting some figures on storage loss. According to a Kobe University study carried out in five countries in Asia, in four of these the main cause of loss was rodents and to a lesser degree insects, which agrees with the Chinese conclusions. A World Bank study of commercial paddy storage in Bangladesh also confirms these figures (12-13 percent of loss caused by insects and rodents, out of a total loss of 19.7 percent). Finally, a study by the Brazilian Technical Commission for Agricultural Loss Reduction divides post-harvest losses into three major headings, giving the following figures:

Harvest	12.6%
Milling	7.0%
Storage	2.4%
Total	22.0%

These broad averages simply give an idea of proportions and are thus of primarily statistical interest. However, it will be noted that, in the Chinese example, storage on its own accounts for more losses than all harvest activities together (harvesting, threshing, drying, cleaning), whereas, in the Brazilian case, storage accounts for less losses than harvest.

Maize

Along with wheat and rice, maize is one of the three major cereals grown throughout the world for human consumption and the total area used for this crop has increased considerably in recent decades. Yields have also risen, thanks to the spread of hybrid varieties and genetic improvements, but it must also be said that these new varieties are biologically more demanding and also, sometimes, physiologically more vulnerable than the old land races, as is often the case with seedlings and seed produced by agricultural research.

However, the main reason for this weakness is the twofold fact that maize is a tropical wetland crop and that at the time of harvesting the grain may have a moisture content of over 30 percent. Immediate harvesting is out of the question unless rapid artificial drying facilities are available, clearly not the case for small farmers in Latin America and Africa, who harvest by hand.

The traditional, economical method is to leave the harvest to dry while standing and wait (for a month and often longer) until the moisture content of the grain has fallen below 15 percent. However, in the rainy season, when the relative air humidity is about 90 percent, as it is in equatorial regions at the time of harvesting, there is no alternative to gathering the damp maize, removing it from the field and drying it under cover.

In any case, it must be borne in mind that the longer the harvest remains standing, the greater the risks of loss: the wind will break dry stalks, so that ears that are too heavy will fall to the ground; rain will encourage the spread of moulds; vertebrates (birds, rodents, monkeys) will take their share, while insects such as the maize weevil (Sitophilus zeamaïs) or the coffee-bean weevil (Araecerus fasciculatus) will lay eggs in the grain; and lastly, many cobs will be eaten or too damaged to be stored or sold, especially in the case of high-yield varieties, which have less developed husks so that the cob is only partially covered.

The study presented at Lomé was carried out in close collaboration with farmers and deserves further attention, not only for its other results but also for its methodological approach. It was a component in a fuller research programme lasting several years and covered seven months' storage (August 1985-March 1986), observing the combined effects of three pre-storage factors (variety of maize, whether or not fertilizer was used and timing of harvesting) on three main groups of parameters: attacks on cobs; loss of dry weight; and yields. Over the seven months, samples were taken every two weeks (making a total of 13) from 24 well-monitored granaries. Depending on the time of harvesting, the moisture content of the maize ranged between 21.1 and 25.1 percent; and again, depending on the time of harvesting, once stored, it took eight to ten weeks to drop to 13 or 14 percent.

The results are summarized as follows:

— average percentages of cobs attacked: these varied between 56.5 and 63.5 percent, with an overall average of 59.8 percent; 10 percent of attacks were due to insects and moulds, but insects were the prime culprits (20 times more damage than moulds); other causes of attack, particularly rodents and birds, were not recorded; (it should be noted that the granaries, set up on the Lomé Agricultural College training farm, were well monitored and protected);

— dry weight losses: these amounted to 9 percent of harvested and stored maize;

— yields: the post-storage yield of the improved variety was 123 percent that of the land race; maize grown with fertilizer gave a yield 131 percent that of maize grown without fertilizer; and early-harvested maize gave a yield 113 percent that of a late-harvested crop.

This is merely a selection of the extensive information supplied by this study, to be taken with due caution. Its interest lies primarily in its concern to explain the method used and the limitations of its work. For example, the following principle is stated with respect to the distinction between early and late harvesting: "A harvest is early when carried out on a field where one stand in five has ears that have tilted downwards; a harvest is late when carried out at least 14 days after the time of an early harvest."

Before comparing these results with those of other studies, it should be recalled that the approach in question had three aspects: (1) the effects of pre-storage factors on traditional maize conservation; (2) preliminary observations on the effects of temperature on the working of traditional granaries; and (3) preliminary observations on the effects of traditional granaries on grain moisture, in function of relative air humidity.

Maize is more difficult to dry than other cereals because of its high moisture content at harvest time and the same problem arises with storage and conservation. Small farmers therefore often prefer to keep the cobs enclosed in their husks in order to protect them against weevils, even if certain varieties have thinner husks and softer grains that make them an easier prey for insects. On the other hand, in regions with a high rainfall, it is better to store cobs without their coverings, so long as they are treated against insects and ensured sufficient aeration. This is traditionally done in the humid tropics and equatorial regions such as the area bordering the Gulf of Benin, where the cobs are carefully arranged in layers on a circular platform and covered with a thick layer of leaves or stubble stacked like a conical hat. The bases of the cobs are turned outwards, so that the gaps between them allow natural ventilation and once the initial drying is completed, this helps to counter rises in temperature and moisture.

However, climates that are regularly moist and warm are bound to encourage the development of insects such as Sitophilus spp. and Prostephanus truncatus that attack harvests and stocks. The latter, known as the larger grain borer, is particularly formidable because of its proliferation and its remarkable appetite. It has gradually spread throughout sub-Saharan Africa from east to west, especially in areas with a high rainfall and has become a veritable plague over the years.

The many surveys carried out in these regions, notably on behalf of FAO, have shown losses from insects of between two and three percent for husked cobs after six months' storage in traditional granaries. The higher losses occurring during threshing as a result of previous damage should then be added to this two and three percent. As growers explain, grain damaged by insects in the course of three months' storage becomes worthless, resulting in a loss of as much as 15 percent. This happens only in regions with no dry season. In more northerly regions with a dry season of about five months (December to April), insects do not appear and proliferate once more until May, when the humidity of the ambient air rises markedly.

Millet and sorghum

Millet and sorghum, which cover a number of botanical families, are short-cycle cereals grown primarily in semi-arid regions. They are usually harvested after the rainy season, so that when they reach maturity there is less danger from humidity than from birds and other field pests, whether rodents or wild or domesticated animals, especially from broken stalks and fallen ears or panicles, not to mention natural husking and thefts. However, there are regions where the harvest is gathered during the rainy season, in which case speedy drying is necessary.

Losses can be serious, depending on how long there is before harvesting and how long harvesting itself takes. They can also be considerable during transport, still often carried out manually. Losses can therefore be heavy, even before the harvest is brought in and stored or sold

Millet is more vulnerable in the field and in transport than sorghum and that early millet is slightly more delicate than late millet. The overall losses for the two operations can reach about five percent for sorghum and over 15 percent, a considerable figure for early millet. In the absence here of figures for losses of late millet during transport, on the basis of on-field losses, it can be estimated that overall losses of late millet would not exceed five or six percent, a figure similar to overall losses for sorghum.

Millet and sorghum do not keep as well as threshed grain. Traditionally, therefore, they are stored in village granaries in ears or panicles; the women then pound them by hand whenever required which is certainly the most economical method of threshing. Flail threshing, which is normally done by a team on a make-shift threshing floor in the village or the fields, pushes much grain into the ground, where it is lost. However, mechanical threshing seems to cause the highest loss because of the length of ears in the case of pearl millet and the density of spikelets on the rachis in the case of other millets and sorghum. Although the many types of pedal or motorized threshing machine now in use save labour and significantly improve yields, they do not prevent breakages and they leave grain on ears and panicles.

If certain tests are to be believed, combine-harvesters can entail heavy losses, between six percent and as much as 30 percent, partially because of the speed at which these machines move. The two figures for threshing losses recorded by the above-cited Gambia project (this time only for millet) are given below and clearly show that losses are higher with mechanical than manual threshing:

Many surveys and studies have been carried out in past years on the methods, conditions and effects of storing cereals, particularly millet and sorghum, in farm or village granaries, traders' storehouses and public or private storehouses, warehouses or silos. They have considered the variety of products, methods and objectives, without forgetting socio-economic and political changes as these affect the agricultural and rural sector, as well as the international market and the agricultural and food trade.

Pulses and oilseeds

Pulses, bean and cowpea, pea, broad bean and lentil in particular, are part of the staple diet of many countries and have the advantage of adding some protein to diets based essentially on cereals, but are harder to harvest and conserve than cereals. When they are mature, the dehiscent pods can open or burst, so that many seeds fall to the ground; this happens during harvesting, but even more so during transport. They are also more vulnerable to insect attack, particularly from weevils, which lay their eggs on the pods or seeds before they are gathered. These pests are often commonly known by the name of their preferred crop, for example, the bean-seed beetle (Acanthoscelides obtectus), the cowpea-seed beetle (Callosobruchus maculatus) and the pea-seed beetle (Zabrotes subfasciatus).

Pulses can have a high moisture content at the time of harvesting, depending on species, variety and climate, as is seen from observations on three varieties of bean (source: the frijoles de la Sierra project in Ecuador).

Effective drying is therefore important and this is generally done by exposure to the sun on terraces or in openwork containers; it will obviously be more effective if the beans have been shelled first. According to the above-mentioned study, the moisture content can drop to 12 or 13 percent after three days in the sun, a level generally sufficient for safe storage, as indicated by the following recommendations.

It can be seen that oilseeds (soybeans and groundnuts) require a lower moisture content than pulses. Shelled groundnuts are especially demanding here, which means that they are especially vulnerable. Unfortunately, even such drying is not enough to provide immunity to parasite attack, particularly from weevils. An ad hoc method of combating this has therefore been further developed and considerably improved by modern technology. In this approach, the container is filled to capacity and then sealed hermetically so that the interstitial air becomes so rarified and inert that it anaesthetizes or even asphyxiates larvae and insects. When larger quantities are involved and the containers or storage space cannot be sealed hermetically, treatment with insecticide powder is recommended. Below are some figures for bean and soybean losses, taken from the national Brazilian study previously cited.

Although soybeans are thus less vulnerable than beans, particularly during storage, the two operations of harvesting and storage account for the majority of losses for the two crops. In the case of beans, storage alone accounts for almost two-thirds of total losses throughout the pipeline, whereas in the case of soybeans, harvesting accounts for almost half the total losses.

Groundnut

Since its fruit grows in the ground, groundnut is different from other oilseeds. The moisture content of the pods is 40 to 50 percent when they are dug up, which has to be reduced to 10 percent before threshing is possible. Natural drying in small heaps can take weeks, during which the insects that have already started their work in the fields will continue their inroads, not to mention the danger of moulds. During storage, Trogoderma will cause more damage than weevils, which normally stay on the surface of the stored groundnuts. Long-term storage therefore necessitates fumigation under tarpaulin or in airtight storehouses.

Roots and tubers

Unlike cereals, tropical root or tuber crops are all fragile, perishable foodstuffs. Harvesting them is a delicate process and must be performed with the utmost care, since it affects all subsequent operations, particularly any action to prevent damage or loss. It is therefore important to avoid cuts and bruising during digging, heaping, gathering and transporting, since such damage leaves the door open to viruses and moulds which inevitably lead to rotting.

Yam

The yam is a tuber crop grown in the humid tropics and also in savannah regions. There are many species, but the white yam or Dioscorea rotundata, which originated in West Africa, is the species most widely grown for human consumption throughout the world.

The tuber reaches maturity after six to nine months' growth and then enters a period of dormancy lasting from four to eighteen weeks, depending on species. Yams can be stored during this period of physiological rest. However, preservation is a delicate matter, since the tuber's high moisture content, between 60 and 80 percent, means that transpiration and respiration continue, even if at a reduced rate. Some figures for the possible ranges of moisture content for the main species of yam will give an idea of the problem here. This is why none of the various storage methods, rudimentary, improved or more sophisticated, can prevent serious losses. The various causes of loss can be placed under five main headings:

— nematodes, primarily the yam nematode (Scutellonema), a particularly dangerous parasite that moves in during the growth period and is therefore present at harvest time;
— rot, caused by moulds and bacteria;
— rodents;
— certain insects, especially the white yam worm and the cochineal;
— tuber budding and respiration, leading to loss of weight and of nutritional reserves and quality; germination often makes yams bitter, while transpiration can result in a weight loss of up to 20 percent in five months of storage.

Cassava

Cassava is a supplementary and security crop in equatorial and tropical regions, held in high regard by the inhabitants. However, it is even harder to keep than yam, because the root starts to deteriorate on the second or third day after harvesting and exposure to the light. For family or local consumption, the simplest natural storage method is therefore to leave the root in the ground after it has matured, digging it up as needed.

Over the years many techniques have been tried in an effort to extend the keeping-time of fresh roots. Some of these, tested particularly in Ghana and Nigeria, have had reasonable results, but the period still does not stretch beyond 15 or 20 days, after which the level of rotting becomes unacceptable (reaching 50 percent after 42 days and 80 percent after 63 days).

Research has therefore focused mainly on processing, in order to improve the farm-level techniques used by women and firstly to detoxify the roots. Fresh cassava contains high levels of hydrocyanic acid, which can lead to poisoning. The initial washing and grating processes are thus intended to detoxify it.

The two main products obtained from the root are cassava flour and gari. In both cases production starts with natural fermentation but the processing operations are much more laborious for gari than flour. The latter, which is easy to store and market, is consumed mainly in the form of foufou, an African staple. Like foufou, gari is sieved after fermentation but instead of being formed into lumps and boiled, it is baked. This process is called garification and produces a couscous-like granular product.

Economic Aspects of Post-harvest Losses

When speaking of food losses, a first consideration is the distinction between perishable and non-perishable or preservable foodstuffs. In one of its studies assessing losses of perishable products through handling or post-harvest operations, the special FAO loss-prevention programme presents the following conclusions: if harvest operations, storing (stacking and/or packaging in warehouse), storage and transport are defective, losses are separated as follows:

Harvesting	5-8 percent
Storing operation	15-20 percent
Storage	5-10 percent
Transport	10-12 percent

i.e. a theoretical total of35-50 percent

Although average losses are not in fact so high when an overall calculation is made, they are still far from negligible. While the situation is certainly better for non-perishable products, it is still unsatisfactory and could be improved. This can be seen, for example, from an Asian study on the post-harvest chain for grains (cereals), which analyses losses from the main operations and estimates percentages for those that could be avoided. Its conclusions are as follows: each year five percent of the losses from poor monitoring and regulation of humidity and another five percent of the losses from problems connected with handling, storage and processing, could be avoided.

In this connection, we would note that in industrialized countries, grain producers accept a one percent wastage during combine-harvesting and a two percent wastage during storage as normal, making three percent for these two operations alone. The current rule is stricter, estimating that an acceptable loss rate would be 0.75 to one percent after nine to ten months' storage. The loss rate accepted in the commercial sector is usually 0.75 percent.

Like loss assessment, calculation or assessment of costs and cost prices entails considerable difficulties, if not uncertainties, if all the relevant factors are to be taken into account. It is particularly important to be clear as to whether the "rational" economy of an industrialized country or that of a developing country is involved. Even then, such figures are always of relative value and are dependent on a socio-economic framework that must never be forgotten. These difficulties are certainly one of the main reasons why economic data are poorly developed in works on post-harvest systems and losses.

Studies on the economics and management of the cereal post-harvest system in France indicate that storage amount comes to 12 percent of the value of the grain. The cost calculations in the summary report on the Accra Workshop cited above have the advantage of comparing the cost/benefit ratio for three different granaries after six months' storage. As before, these figures refer to trials on maize storage in Benin, i.e. a small-scale, village environment and compare the performance of a number of improved granaries with that of a control granary of the traditional type. The maize was stored in cobs still in their husks and the two improved granaries, BT 2 and BT 3, had capacities of two and six tonnes, respectively. The document specifies that "the cost of storage comprises costs of construction, phytosanitary treatment and handling into storage" and that "the cost/benefit ratio is the greatest value achieved by delaying the sales of maize".

References

Bourne, M.C. 1977. *Post-harvest Food Losses - The Neglected Dimension in Increasing the World Food Supply.* New York, Cornell University International Agriculture.

Boxall, R.A. 1986. *A Critical Review of the Methodology for Assessing Farm-level Grain Losses after Harvest.* Londres, TDRI, G191, 139 p.

Calverley, D.J.B. 1996. A *study of loss assessment in eleven projects in Asia concerned with rice.* Rome, FAO.

Harris, K.L., Lindblad, C.J. & Al. 1978. *Post-harvest Grain Loss Assessment Methods.* Washington & Londres.

NRI. 1994. *Training Manuel on Grains.* Rome, FAO.

2

Harvesting and Preparation for Market

The three main objectives of applying postharvest technology to harvested fruits and vegetables are:

1. to maintain quality (appearance, texture, flavor and nutritive value)
2. to protect food safety, and
3. to reduce losses between harvest and consumption.

Effective management during the postharvest period, rather than the level of sophistication of any given technology, is the key in reaching the desired objectives. While large scale operations may benefit from investing in costly handling machinery and high-tech postharvest treatments, often these options are not practical for small-scale handlers. Instead, simple, low cost technologies often can be more appropriate for small volume, limited resource commercial operations, farmers involved in direct marketing, as well as for suppliers to exporters in developing countries.

Many recent innovations in postharvest technology in developed countries have been in response to the desire to avoid the use of costly labor and the desire for cosmetically "perfect" produce. These methods may not be sustainable over the long term, due to socioeconomic, cultural and/or environmental concerns. For example, the use of postharvest pesticides may reduce the incidence of surface defects but can be costly both in terms of money and environmental consequences. In addition, the growing demand for organically produced fruits and vegetables offers new opportunities for small-scale producers and marketers.

Local conditions for small-scale handlers may include labor surpluses, lack of credit for investments in postharvest technology, unreliable electric power supply, lack of transport options, storage facilities and/or packaging materials, as well as a host of other constraints. Fortunately, there is a wide range of simple postharvest technologies from which to choose, and many practices have the potential of meeting the special needs of small-scale food handlers and marketers. Many of the practices included in the manual have successfully been used to reduce losses and maintain produce quality of horticultural crops in various parts of the world for many years.

Mechanical damage during harvest can become a serious problem, as injuries predispose produce to decay, increased water loss and increased respiratory and ethylene production rates leading to quick deterioration. In general, harvesting by machine will cause more damage than harvesting by hand, although some root crops can be severely damaged by careless hand digging.

The containers used by pickers in the field should be clean, have smooth inside surfaces and be free of rough edges. Stackable plastic crates, while initially expensive, are durable, reusable and easily cleaned. If baskets are used, they should be woven "inside out" with the stubs of the beginning and end of each cane on the outside of the basket.

Manual harvesters should be well trained in the proper way to harvest the crop to minimize damage and waste, and should be able to recognize the proper maturity stage for the produce they are handling. Pickers should harvest with care, by snapping, cutting or pulling the fruit or vegetable from the plant in the least damaging manner. The tips of knives should be rounded to minimize inadvertent gouges and excess damage to perennial plants. Knives and clippers should always be well sharpened. Pickers should be trained to empty their picking bags and/or baskets with care, never dumping or throwing produce into field containers.

If harvesters pick directly into large bulk bins, produce can be protected from bruising by the use of a deaccelerating chute fashioned from canvas. Vented, stackable field containers should be kept clean and smooth.

Exposure to the sun should be avoided as much as possible during and after harvest, as produce left out in the sun will gain heat and may become sun-burned. Field bins should be placed in the shade or loosely covered (for example with light colored canvas, leafy plant materials, straw or an inverted empty container) if delays are expected in removing them from the field. Night or early morning harvest is sometimes an option for harvesting produce when internal temperatures are relatively low, reducing the energy needed for subsequent cooling. Latex flow is often lower later in the morning than it is at dawn for crops such as mango and citrus, so harvesting as late in the morning as possible can also reduce later efforts required to clean the produce before packing.

Directly following harvest, when produce is prepared for marketing, cooling is essential. Cooling (also known as "pre-cooling") is the removal of field heat directly after harvest, before any further handling. Any delays in cooling will shorten postharvest life and reduce quality. Even produce undergoing repeated cooling and warming deteriorates at a slower rate than produce that has not been cooled.

Rough handling during preparation for market will increase bruising and mechanical damage and limit the benefits of cooling. Roads between the field and the packinghouse should be graded and free from large ruts, bumps and holes. Field boxes must be well-secured during transport and, if stacked, not overfilled. Transport speeds must be suited to the quality and conditions of the roads, and truck and/or trailer suspensions kept in good repair. Reduced tire air pressure on transport vehicles will reduce the amount of motion transmitted to the produce..

Any practice that reduces the number of times the produce is handled will help reduce losses. Field packing (selection, sorting, trimming and packaging of produce at the time of harvest) can greatly reduce the number of handling steps the produce must undergo before marketing. Small, mobile field packing stations can be designed to be moved along with the packers and to provide shade for packing operations.

Maturity Standards

Maturity standards have been determined for many fruit, vegetable and floral crops. Harvesting crops at the proper maturity allows handlers to begin their work with the best possible quality produce. Produce harvested too early may lack flavor and may not ripen properly, while produce harvested too late may be fibrous or overripe. Pickers can be trained in methods of identifying produce that is ready for harvest. The following table, from Reid provides some examples of maturity indices.

Table 1. Examples of maturity indices

Index	*Examples*
Elapsed days from full bloom to harvest	Apples, pears
Mean heat units during development	Peas, apples, sweet corn
Development of abscission layer	Some melons, apples, feijoas
Surface morphology and structure	Cuticle formation on grapes, tomatoes Netting of some melons Gloss of some fruits (development of wax)
Size	All fruits and many vegetables
Specific gravity	Cherries, watermelons, potatoes
Shape	Angularity of banana fingers Full cheeks of mangos Compactness of broccoli and cauliflower
Solidity	Lettuce, cabbage, brussels sprouts
Textural properties	
Firmness	Apples, pears, stone fruits
Tenderness	Peas
Color, external	All fruits and most vegetables
Internal color and structure	Formation of jelly-like material in tomato fruits Flesh color of some fruits
Compositional factors	
Starch content	Apples, pears
Sugar content	Apples, pears, stone fruits, grapes
Acid content, sugar/acid ratio	Pomegranates, citrus, papaya, melons, kiwifruit
Juice content	Citrus fruits
Oil content	Avocados
Astringency (tannin content)	Persimmons, dates
Internal ethylene concentration	Apples, pears

Vegetables are harvested over a wide range of maturities, depending upon the part of the plant used as food. Table 2 provides some examples of maturity indices of vegetable crops.

Maturity Indices for Fruits and Vegetables

Ripening follows or overlaps maturation, rendering the produce edible, as indicated by taste. Senescence is the last stage, characterised by natural degradation of the fruit or vegetable, as in loss of texture, flavour, etc. (senescence ends at the death of the tissue of the fruit). Some typical maturity indexes are described below:

Skin colour

This factor is commonly applied to fruits, since skin colour changes as fruit ripens or matures. Some fruits exhibit no perceptible colour change during maturation, depending on the type of fruit or vegetable. Assessment of harvest maturity by skin colour depends on the judgment of the harvester, but colour charts are available for cultivars, such as apples, tomatoes, peaches, chilli peppers, etc.

Table 2. Examples of maturity indices of vegetable crops

Root, bulb and tuber crops	
Radish and carrot	Large enough and crispy (overmature if pithy)
Potato, onion, and garlic	Tops beginning to dry out and topple down
Yam bean and ginger	Large enough (overmature if tough and fibrous)
Green onion	Leaves at their broadest and longest
Fruit vegetables	
Cowpea, yard-long bean, snap bean, batao, sweet pea, and winged bean	Well-filled pods that snap readily
Lima bean and pigeon pea	Well-filled pods that are beginning to lose their greenness
Okra	Desirable size reached and the tips of which can be snapped readily
Upo, snake gourd, and dishrag gourd	Desirable size reached and thumbnail can still penetrate flesh readily (ove-rmature if thumbnail cannot penetrate flesh readily)
Eggplant, bitter gourd, chayote or slicing cucumber	Desirable size reached but still tender (over-mature if color dulls or changes and seeds are tough)
Sweet corn	Exudes milky sap when thumbnail penetrates kernel
Tomato	Seeds slipping when fruit is cut, or green color turning pink
Sweet pepper	Deep green color turning dull or red
Muskmelon	Easily separated from vine with a slight twist leaving clean cavity
Honeydew melon	Change in fruit color from a slight greenish white to cream; aroma noticeable
Watermelon	Color of lower part turning creamy yellow, dull hollow sound when thumped
Flower vegetables	
Cauliflower	Curd compact (over-mature if flower cluster elongates and become loose)
Broccoli	Bud cluster compact (over-mature if loose)
Leafy vegetables	
Lettuce	Big enough before flowering
Cabbage	Head compact (over-mature if head cracks)
Celery	Big enough before it becomes pithy

Optical methods

Light transmission properties can be used to measure the degree of maturity of fruits. These methods are based on the chlorophyll content of the fruit, which is reduced during maturation. The fruit is exposed to a bright light, which is then switched off so that the fruit is in total darkness. Next, a sensor measures the amount of light emitted from the fruit, which is proportional to its chlorophyll content and thus its maturity.

Shape

The shape of fruit can change during maturation and can be used as a characteristic to determine harvest maturity. For instance, a banana becomes more rounded in cross-sections and less angular as it develops on the plant. Mangoes also change shape during maturation.

As the mango matures on the tree the relationship between the shoulders of the fruit and the point at which the stalk is attached may change. The shoulders of immature mangoes slope away from the fruit stalk; however, on more mature mangoes the shoulders become level with the point of attachment, and with even more maturity the shoulders may be raised above this point.

Size

Changes in the size of a crop while growing are frequently used to determine the time of harvest. For example, partially mature cobs of Zea mays saccharata are marketed as sweet corn, while even less mature and thus smaller cobs are marketed as baby corn. For bananas, the width of individual fingers can be used to determine harvest maturity. Usually a finger is placed midway along the bunch and its maximum width is measured with callipers; this is referred to as the calliper grade.

Aroma

Most fruits synthesise volatile chemicals as they ripen. Such chemicals give fruit its characteristic odour and can be used to determine whether it is ripe or not. These doors may only be detectable by humans when a fruit is completely ripe, and therefore has limited use in commercial situations.

Fruit opening

Some fruits may develop toxic compounds during ripening, such as ackee tree fruit, which contains toxic levels of hypoglycine. The fruit splits when it is fully mature, revealing black seeds on yellow arils. At this stage, it has been shown to contain minimal amounts of hypoglycine or none at all. This creates a problem in marketing; because the fruit is so mature, it will have a very short post-harvest life.

Analysis of hypoglycine 'A' (hyp.) in ackee tree fruit revealed that the seed contained appreciable hyp. at all stages of maturity, at approximately 1000 ppm, while levels in the membrane mirrored those in the arils. This analysis supports earlier observations that unopened or partially opened ackee fruit should not be consumed, whereas fruit that opens naturally to over 15 mm of lobe separation poses little health hazard, provided the seed and membrane portions are removed. These observations agree with those of Brown et al. who stated that bright red, full sized ackee should never be forced open for human consumption.

Leaf changes

Leaf quality often determines when fruits and vegetables should be harvested. In root crops, the condition of the leaves can likewise indicate the condition of the crop below ground. For example, if potatoes are to be stored, then the optimum harvest time is soon after the leaves and stems have died. If harvested earlier, the skins will be less resistant to harvesting and handling damage and more prone to storage diseases.

Abscission

As part of the natural development of a fruit an abscission layer is formed in the pedicel. For example, in cantaloupe melons, harvesting before the abscission layer is fully developed results in inferior flavoured fruit, compared to those left on the vine for the full period.

Firmness

A fruit may change in texture during maturation, especially during ripening when it may become rapidly softer. Excessive loss of moisture may also affect the texture of crops. These textural changes are detected by touch, and the harvester may simply be able to gently squeeze the fruit and judge whether the crop can be harvested. Today sophisticated devices have been developed to measure texture in fruits and vegetables, for example, texture analysers and pressure testers; they are currently available for fruits and vegetables in various forms.

A force is applied to the surface of the fruit, allowing the probe of the penetrometer or texturometer to penetrate the fruit flesh, which then gives a reading on firmness. Hand held pressure testers could give variable results because the basis on which they are used to measure firmness is affected by the angle at which the force is applied. Two commonly used pressure testers to measure the firmness of fruits and vegetables are the Magness-Taylor and UC Fruit Firmness testers (Figure 1). A more elaborate test, but not necessarily more effective, uses instruments like the Instron Universal Testing Machine. It is necessary to specify the instrument and all settings used when reporting test pressure values or attempting to set standards.

Juice content

The juice content of many fruits increases as the fruit matures on the tree. To measure the juice content of a fruit, a representative sample of fruit is taken and then the juice extracted in a standard and specified manner.

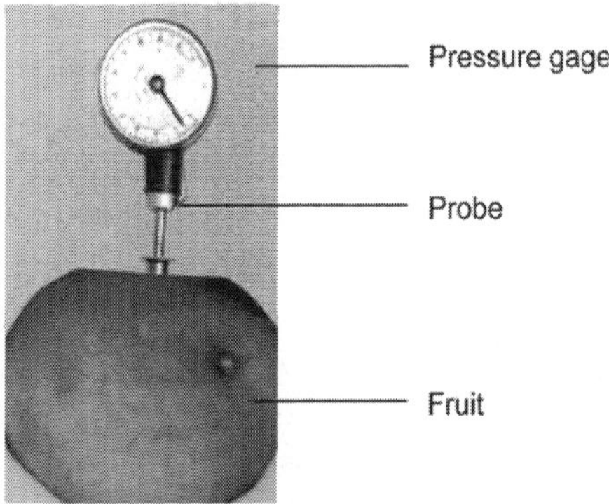

Figure 1. Pressure tester used to measure firmness of fruits and vegetables.

The juice volume is related to the original mass of juice, which is proportional to its maturity. The minimum values for citrus juices are presented in Table 1.

Table 1. Minimum juice values for mature citrus.

Citrus fruit	*Minimum juice content (%)*
Naval oranges	30
Other oranges	35
Grapefruit	35
Lemons	25
Mandarins	33
Clementines	40

Oil content and dry matter percentage

Oil content can be used to determine the maturity of fruits, such as avocados. According to the Agricultural Code in California, avocados at the time of harvest and at any time thereafter, shall not contain in weight less than 8% oil per avocado, excluding skin and seed. Thus, the oil content of an avocado is related to moisture content. The oil content is determined by weighing 5-10 g of avocado pulp and then extracting the oil with a solvent in a destillation column. This method has been successful for cultivars naturally high in oil content.

A round flask is used for the solvent. Heat is supplied with an electric plate and water recirculated to maintain a constant temperature during the extraction process. Extraction is performed using solvents such as petroleum ether, benzene, diethyl ether, etc., a process that takes between 4-6 h. After the extraction, the oil is recovered from the flask through evaporation of the water at 105°C in an oven until constant weight is achieved.

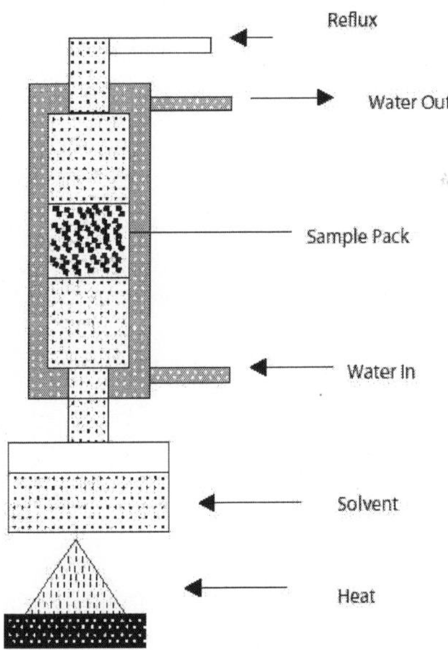

Figure 2. Distillation column used for oil determination.

Moisture content

During the development of avocado fruit the oil content increases and moisture content rapidly decreases. The moisture levels required to obtain good acceptability of a variety of avocados cultivated in Chile are listed in Table 3.

Table 3. Moisture content of avocado fruit cultivated in Chile

Cultivar	*Moisture content (%)*
Negra de la Cruz	80.1
Bacon	77.5
Zutano	80.5
Fuerte	77.9
Edranol	78.1
Hass	73.8
Gwen	78.4
Whitesell	79.1

Sugars

In climacteric fruits, carbohydrates accumulate during maturation in the form of starch. As the fruit ripens, starch is broken down into sugar. In non-climacteric fruits, sugar tends to accumulate during maturation. A quick method to measure the amount of sugar present in fruits is with a brix hydrometer or a refractometer. A drop of fruit juice is placed in the sample holder of the refractometer and a reading taken; this is equivalent to the total amount of soluble solids or sugar content. This factor is used in many parts of the world to specify maturity. The soluble solids content of fruit is also determined by shining light on the fruit or vegetable and measuring the amount transmitted.

Starch content

Measurement of starch content is a reliable technique used to determine maturity in pear cultivars. The method involves cutting the fruit in two and dipping the cut pieces into a solution containing 4% potassium iodide and 1% iodine. The cut surfaces stain to a blue-black colour in places where starch is present. Starch converts into sugar as harvest time approaches. Harvest begins when the samples show that 65-70% of the cut surfaces have turned blue-black.

Acidity

In many fruits, the acidity changes during maturation and ripening, and in the case of citrus and other fruits, acidity reduces progressively as the fruit matures on the tree. Taking samples of such fruits, and extracting the juice and titrating it against a standard alkaline solution, gives a measure that can be related to optimum times of harvest. Normally, acidity is not taken as a measurement of fruit maturity by itself but in relation to soluble solids, giving what is termed the brix: acid ratio.

Specific gravity

Specific gravity is the relative gravity, or weight of solids or liquids, compared to pure distilled water at 62°F (16.7°C), which is considered unity. Specific gravity is obtained by comparing the weights of equal bulks of other bodies with the weight of water. In practice, the fruit or vegetable is weighed in air, then in pure water.

The weight in air divided by the weight in water gives the specific gravity. This will ensure a reliable measure of fruit maturity. As a fruit matures its specific gravity increases. This parameter is rarely used in practice to determine time of harvest, but could be used in cases where development of a suitable sampling technique is possible. It is used however to grade crops according to different maturities at post-harvest. This is done by placing the fruit in a tank of water, wherein those that float are less mature than those that sink.

Harvesting Practices

Harvesting practices should cause as little mechanical damage to produce as possible. Gentle digging, picking and handling will help reduce crop losses.

For some crops, a natural break point forms at the junction of the stem and the stalk when produce is mature. Harvesters should grasp the product firmly but gently and pull upward as illustrated below. Wearing cotton gloves, trimming fingernails, and removing jewelry such as rings and bracelets can help reduce mechanical damage during harvest.

If a small amount of leafy vegetables are being harvested for home use or for sale at a nearby roadside or farmers' market, a small tub of cold water can be useful for cooling the produce. The tub can be brought directly to the field and used by the picker as a field container. Clean water should be used with each lot of produce. Chilling leafy vegetables by using cold water at harvest will help maintain quality and prevent wilting.

Picking baskets, bags and buckets come in many sues and shapes. These harvesting containers can be made by sewing bags with openings on both ends, fitting fabric over the open bottom of ready-made baskets, fitting bags with adjustable harnesses, or by simply adding some carrying straps to a small basket.

Plastic crates are relatively expensive but are durable, reusable and easy to clean. When empty, they can be nested to save space in storage or transport.

Tools Used for Harvesting

Some fruits need to be clipped or cut from the parent plant. Clippers or knives should be kept well sharpened.

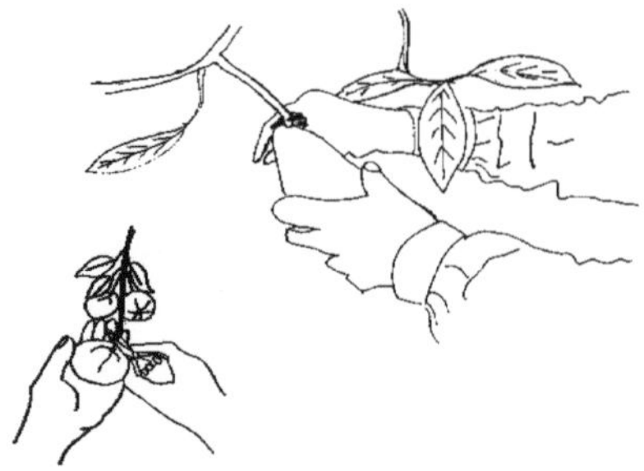

Figure 3. Clipping fruit from plant

Penduncles, woody stems or spurs should be trimmed as close as possible to prevent fruit from damaging neighboring fruits during transport. Pruning shears are often used for harvesting fruits, some vegetables, and cut flowers. A variety of styles are available as hand held or pole models, including shears that cut and hold onto the stem of the cut product. This feature allows the picker to harvest without a catching bag and without dropping fruits.

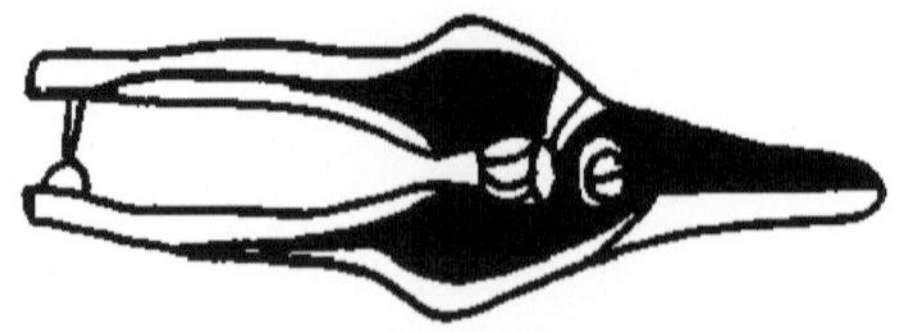

Figure 4. Straight bladed hand shears for fruits and flowers

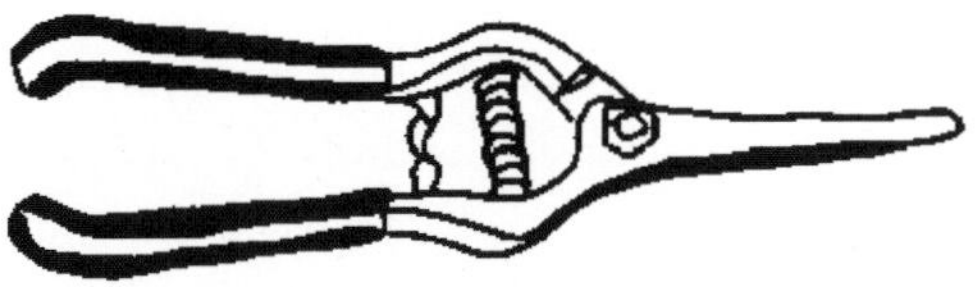

Figure 5. Thin curved blade for grapes and fruits

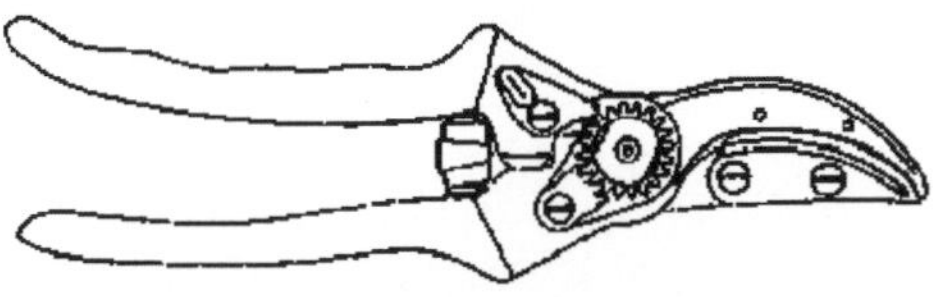

Figure 6. Cut and hold hand shears

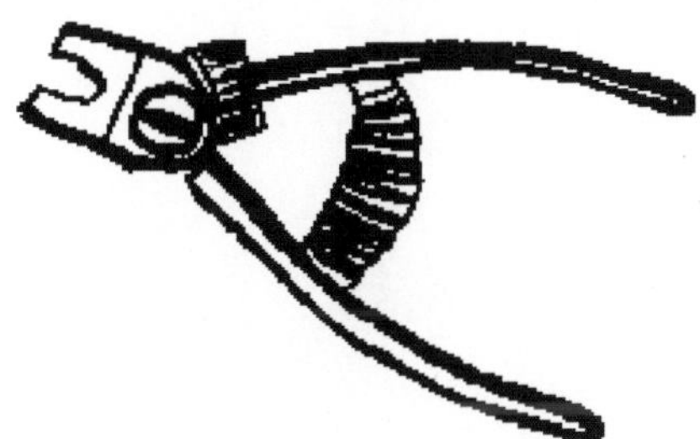

Figure 7. Clipper for citrus fruits

Figure 8. Pole mounted cut and hold picking shears:

Using a cutting tool attached to a long pole can aid picking of crops such as mangoes and avocados when the fruit is difficult to reach. Cutting edges should be kept sharpened and the catching bag should be relatively small The angle of the cutting edge and the shape of the catching bag can affect the quality of the fruit harvested, so it is important to check performance carefully before using any new tools.

Picking poles and catching sacks can be made by hand or purchased from horticultural supply companies. The collection bags illustrated below were hand woven from strong cord or sewn from canvas. The hoop used as the basket rim and cutting edges can be fashioned from sheet metal, steel tubing or recycled scrap metal.

Fruit trees are sometimes quite tall and letting fruit fall to the ground when it is cut from the tree will cause severe bruising. If two pickers work together, one can clip or cut the fruit from the tree, and the other can use a sack to break its fall. The catcher supports the bag with his hands and one foot, catches the falling fruit, then lowers the far end of the bag to allow the fruit to roll safely to the ground.

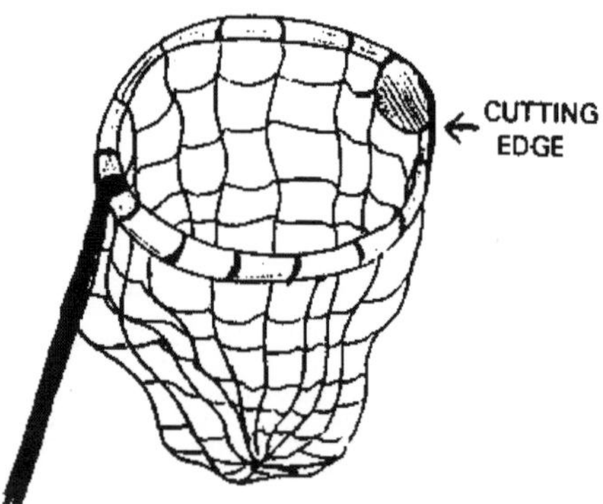

Figure 9. Hand woven collection bag

Unlike most nut crops, pistachio nuts should not be knocked to the ground during harvest because of their open shells and relatively high moisture content. The harvesting practice illustrated below can be used with pistachios and olives with good results. Plastic sheeting or canvas tarpaulins are spread below the tree being harvested, and trees are mechanically shaken or hand knocked (the branches hit with mauls) until the nuts drop. In the illustration below, two harvesters are gathering a sheet covered with produce.

Field Packing

When crops are field packed the picker harvests and then immediately packs the produce after minimal handling Strawberries are generally field packed, since even a small amount of handling will damage these soft fruits. When lettuce is field packed, several wrapper leaves are left on the head to help cushion the produce during transport. A small cart can help reduce the amount of bending and lifting the picker has to do during harvest. The carts shown below have a single wheel in front, and can be pushed along the row ahead the e picker.

Figure 10. Field packing lettuce

A simple aid for field packers is a movable cart with a rack for boxes and a roof to provide shade. This cart is designed to be pushed along the outer edge of the small field where harvest is taking place.

Figure 11. Cart for field packing

This cart for field packing is designed to be pulled by a small tractor into the field when the crop is harvested. This type of cart can be used for field packing many types of crops. The roof folds down for easy transport, and opens up to provide a wide area of shade for the packers and the commodity. The cart design can be modified as needed to suit various products and different operations.

Transport to the Packinghouse

When crops are harvested at some distance from the packinghouse, the produce must be transported before packing. The gravity driven conveyor system for bananas illustrated below provides an example of how handling can be minimised during preparation for market. Harvested bananas are carried to the platforms set up along the conveyor route, then lifted and hung from hooks attached to the wire. Transport speed is controlled by workers who lead the produce to the packinghouse at the bottom of the hill.

Polyethylene Bags

Clear polyethylene bags are used to pack banana bunches in the field, which are then transported to the packinghouse by means of mechanical cableways running through the banana plantation. This technique of packaging and transporting bananas reduces damage to the fruit caused by improper handling.

Plastic Field Boxes

These types of boxes are usually made of polyvinyl chloride, polypropylene, or polyethylene. They are durable and can last many years. Many are designed in such a way that they can nest inside each other when empty to facilitate transport, and can stack one on top of the other without crushing the fruit when full (Figure 12).

Figure 12. Plastic field boxes with nest/stack design.

Wooden Field Boxes

These boxes are made of thin pieces of wood bound together with wire. They come in two sizes: the bushel box with a volume of 2200 in^3 (36052 cm^3) and the half-bushel box. They are advantageous because they can be packed flat and are inexpensive, and thus could be non-returnable.

Figure 13. Typical wooden crate holding fresh tomatoes.

They have the disadvantage of providing little protection from mechanical damage to the produce during transport. Rigid wooden boxes of different capacities are commonly used to transport produce to the packinghouse or to market.

Bulk Bins

Bulk bins of 200-500 kg capacity are used for harvesting fresh fruits and vegetables. These bins are much more economical than the field boxes, both in terms of fruit carried per unit volume and durability, as well as in providing better protection to the product during transport to the packinghouse. They are made of wood and plastic materials. Dimensions for these bins in the United States are 48 × 40 in, and 120 × 100 cm in metric system countries. Approximate depth of bulk bins depends on the type of fruit or vegetable being transported (Table 3).

Table 3. Approximate depth of bulk bins.

Commodity	Depth (cm)
Citrus	70
Pears, apples	50
Stone fruits	50
Tomatoes	40

REFERENCES

Bachmann, J. and R. Earles. 2000. Postharvest Handling of Fruits and Vegetables. *ATTRA Horticulture Technical* Note. 19 pp. (http://www.attra.ncat.org)

FAO. 1986. *Improvement of Post-Harvest Fresh Fruits and Vegetables Handling. Regional Office for Asia and the Pacific*. Maliwan Mansion, Phra Atit Road, Bangkok, 10200, Thailand.

Campbell-Platt, G. 1987. *Fermented Foods of the World: Dictionary and Guide.* Stonam, Massachussetts: Butterworth Heineman.

Zagory, D. 1999. Effects of post-processing handling and packaging on microbial populations. *Postharv. Biol. Technol*. 15:313-321.

3

Curing of Food Crops

Even after they have matured and are removed from the plants, fruits and vegetables are still alive. The objective of curing and storing them is to prolong the post-harvest life of the products. When mature, winter squashes such as Butternut, Acorn, and Hubbard types have hard skins that resist puncture from your thumbnail. Skins of winter squash appear dull and dry compared to the fresh, bright sheen of the skins of immature fruit. Be sure to leave a long stem (handle) on pumpkins, but on winter squashes such as the Hubbard types, be sure to remove the stems completely.

Curing involves elevating storage temperatures to 80 to 85 degrees F with 75 to 80 percent relative humidity for approximately 10 days. Curing heals wounds, helps ripen immature fruit, enhances color, and ensures a longer post-harvest life. After curing, reduce the temperature and relative humidity. Curing is beneficial to pumpkins and some winter squash, but Butternut, Hubbard, and Quality squashes have not shown any added benefits from curing.

Curing root and tuber crops such as sweetpotatoes, potatoes, cassava and yarns is an important practice if these crops are to be stored for any length of time. Curing is accomplished by holding the produce at high temperature and high relative humidity for several days while harvesting wounds heal and a new, protective layer of cells form. While curing can be initially costly, the long extension of storage life makes the practice economically worthwhile.

The best conditions for curing vary among crops as shown in the following table:

Commodity	*Temperature*		*Relative Humidity*	*Days*
	C	F	(%)	
Potato	15-20	59-68	90-95	5-10
Sweetpotato	30-32	86-90	85-90	4-7
Yarns	32-40	90-104	90-100	1-4
Cassava	30-40	86-104	90-95	2-5

Curing, when used for onions, garlic and flowering bulbs refers to the practice directly following harvest, of allowing the external layers of skin and neck tissue to dry out prior to handling and storage. If local weather conditions permit, these crops can be undercut in the field, windrowed and left there to dry for five to ten days. The dried tops of the plants can be arranged to cover and shade the bulbs during the curing process, protecting the produce from excess heat and sunburn. If forced heated air is used for curing onions and other bulbs, one day or less at 35 to 45 C (95 to 113 F) and 60 to 75% relative humidity is recommended. The dried layers of 'skin. then protect the produce from further water loss during storage.

Field Curing

Yams and other tropical root and tuber crops can be cured outdoors if piled in a partially shaded area. Cut grasses or straw can be used as insulating materials and the pile should be covered with canvas, burlap or woven grass mats. Curing requires high temperature and high relative humidity, and this covering will trap self-generated heat and moisture. The stack should be left for about four days.

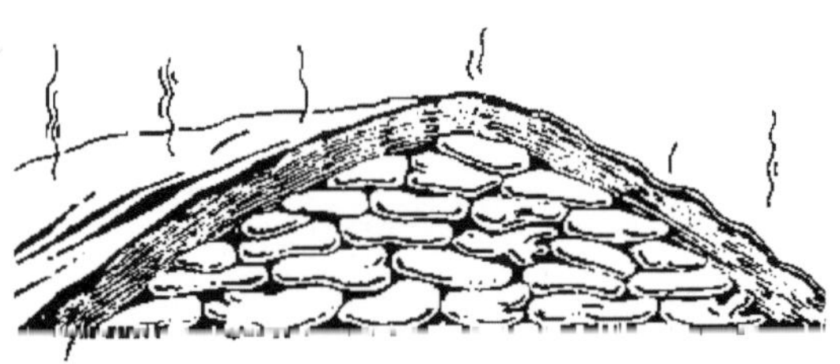

Figure 1. Cut-away view of Yam Curing

Onions and garlic can be cured in the field in regions where harvest coincides with the dry season. The crops can be cured either in windrows or after packing into large fiber or net sacks. The produce can be left in the field for five days, then checked daily until the outer skin and neck tissues are properly dried. Curing may take up to ten days, depending on weather conditions.

Curing can be assisted by the use of ventilated sheds in regions where solar radiation and/or relative humidity is high or natural air movement is low. Produce in sacks can be stacked in the shade on canvas tarpaulins under one or more ceiling fans.

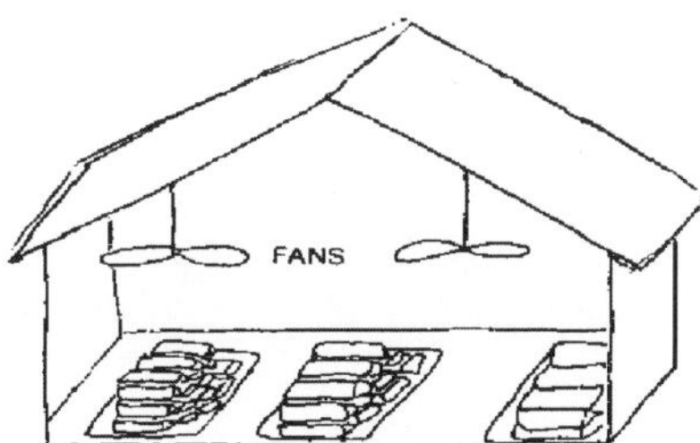

Figure 2. Curing assisted by shade and ventilation

CURING WITH HEATED AIR

The most uniform distribution of heat is obtained when heat is introduced near the floor level of a curing structure. Heaters can be placed on the floor near the bins of produce, or heat can be ducted in from outside the curing room. A high relative humidity can be obtained by wetting the floor or by using an evaporative cooler in the room without introducing outside air.

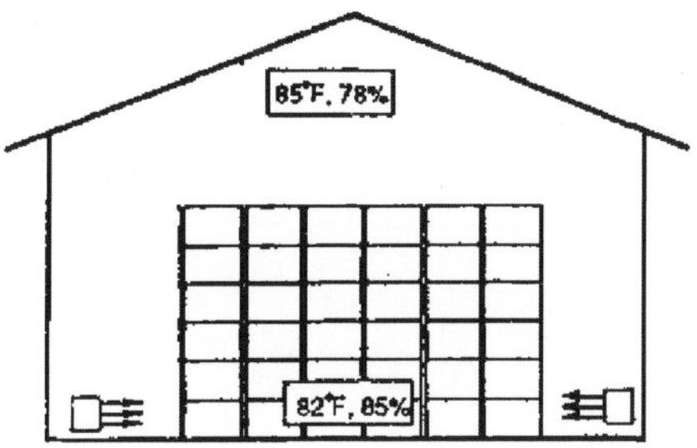

Figure 3.

If heaters are located near the ceiling, then ceiling fans can be used to help redistribute the heat down into the room of produce. Bulk bins must be stacked to allow a gap of 10 to 15 cm (4 to 6 inches) between rows for adequate air circulation.

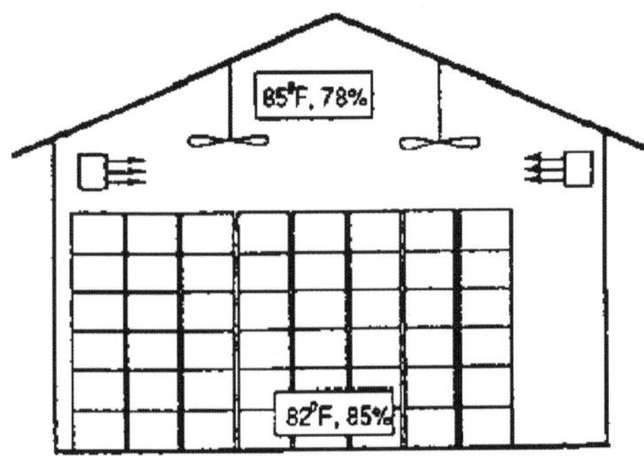

Figure 4.

BULK SYSTEMS FOR CURING ONIONS

Curing using a bulk system requires a fan, a heating unit and a slatted floor. The illustrations below shows how air can be brought in, heated and distributed through a load of bulk onions in the curing room An exhaust opening near the ceiling recirculates heated air.

When using heated air it is easy to overdry the bulbs, leading to loss of external scales and exposure of the fleshy scales underneath. Curing onions should be checked regularly to avoid overdrying.

EMERGENCY CURING

If conditions such as rain or flooded fields do not permit field curing and curing facilities are not available, a temporary tent can be used for curing onions. In the example illustrated below, the tent is constructed from large tarps. Heated air is forced into a hollow area (known as a plenum) at the center of the bins of produce. Several fans are used to circulate the warm air through the onions while they are curing.

REFERENCES

Fellows, P. and Hampton, A 1992. *Small-Scale Food Processing A Guide to Appropriate Equipment*. London: Intermediate Technology Publications. 158 pp.

Hardenburg, R.E.. et al. 1986. *The Commercial Storage of Fruits. Vegetables*. and Florist and Nursery Stocks. USDA Agriculture Handbook 66. 130 pp.

Harvey, E. et al. 1990. Harvesting and postharvest handling of papayas in the Caribbean. Bridgetown, Barbados: Inter-American Institute for Cooperation on Agriculture (IICA).

Lurie, S. 1998. Postharvest heat treatments of horticultural crops. *Hort. Rev.* 22:91-121.

4

Packinghouse Operations

Packinghouse operations can be as simple as moving produce from a field lug into a shipping container, or may include a variety of handling practices, from cleaning, waxing, sizing, and quality grading to color sorting. The provision of shade during the packing operations is extremely important. Shade can be created using palm leaf fronds, a plastic mesh or canvas sheet hung from temporary poles, or via a permanent roofed structure. When deciding upon where to locate a packinghouse, access to the field and market point, adequate space for vehicles to enter and leave the packinghouse and ease of access to labor will all be considerations.

In the simplest packinghouse, produce is delivered in picking containers, immediately after harvest, directly to the packers. The packers then sort, grade, size and pack the produce directly into appropriate transport containers. In this case, each worker must be knowledgeable regarding produce defects, grade and size requirements, and packing methods.

As the size and complexity of the packinghouse increases, more operations and workers trained in specific tasks might be added.

Produce must somehow be removed from the field bin or harvesting container and moved through the packinghouse. This first step is known as ''dumping". Dumping must be done gently, whether using water assisted methods or dry dumping. Wet dumping can decrease bruising and abrasions by using moving, chlorinated (100-150 ppm) water to carry delicate produce. When using dry dumping, padded, sloped ramps or moving conveyor belts can decrease injuries to produce.

Pre-sorting produce is usually done to eliminate injured, decayed, or otherwise defective produce (culls) before cooling or additional handling. Pre-sorting will save energy in that culls will not be handled. Removing decaying produce items will limit the spread of infection to other units, especially if postharvest pesticides are not being used.

For some commodities, such as kiwifruits and avocadoes, dry brushing may be sufficient to clean the produce. Other commodities, however, such as bananas and carrots, require washing. The choice of brushing and/or washing will depend upon both the type of commodity and the type of contamination.

Sanitation is essential, both to control the spread of disease from one item to another, and to limit spore buildup in wash water or in the packinghouse air. Chlorine treatments (100 to 150 ppm Cl) can be used in wash water to help control pathogen buildup during packing operations. There is some variation in the strength of bleach available commercially in different countries, but a rule of thumb is to use 1 to 2 mls of chlorine bleach per liter (1 to 2 ounces of chlorine bleach per 8 gallons of clean water). Walls, floors and packing equipment can also be cleaned using quarternary ammonium compounds labelled as safe for food processing equipment.

Waxing of immature fruit vegetables such as cucumbers and summer squash; mature fruit vegetables such as eggplant, peppers and tomatoes; and fruits such as apples and peaches is common. Food grade waxes are used to replace some of the natural waxes removed in washing and cleaning operations, and can help reduce water loss during handling and marketing. If produce is waxed, the wax coating must be allowed to dry thoroughly before further handling.

Sizing produce is optional but may be worthwhile if certain size grades receive a higher price than others. In most low-input packinghouses, manual sizing is still commonly practiced. Operators should be trained in selecting the size desired and to either directly pack the items into containers or place the selected produce gently into a bin for packing further down the line. Sizing can be done subjectively (visually) with the use of standard size guages. Examples of the smallest and largest acceptable sizes for each product can be placed within view of the operator for easy reference. Hand held sizers are used for a variety of products.

Several types of mechanical sizers are available for small scale operations. One type is composed of a long slanted tray with a series of openings which converge (largest at the top, smallest at the bottom). This type of sizer works best with round commodities. Other sizers are designed as conveyors fitted with chain or plastic belts with various sized openings, and are useful for sizing most commodities. Another simple method for mechanical sizing is to use a set of diverging bar rollers, where the smallest sized produce falls through the rollers first to a sorting belt or bin, and larger sized produce falls between successively more divergent rollers.

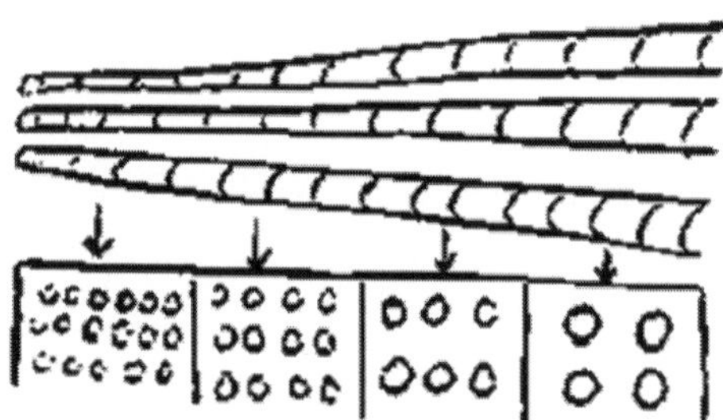

Figure 1. Diverging bar rollers sizer

The typical series of operations in a packinghouse are illustrated below. Dumping can be done using either dry or water-assisted methods, depending upon the sort of produce being handled. Cleaning, as well, can be by washing with chlorinated water or dry brushing alone. Waxing, if practiced, occurs after washing and removal of surface moisture. Grading, as illustrated, separates the product into processing and fresh market categories. Sizing further separates the product, with the smallest size going to the local market or to processing. Typically, the best quality produce is packaged and marketed at the regional or national level.

Dumping

Any time produce is dumped from one container into another, care should be taken to reduce mechanical damage to the commodity. When dumping produce from field bins or from transport vehicles into the packinghouse, dry or wet dumping can be practiced. When using dry dumping practices, the filed container should be emptied slowly and gently onto a tilted ramp with padded edges. In the illustration below, a conveyor belt then carries the dry dumped produce into the packinghouse.

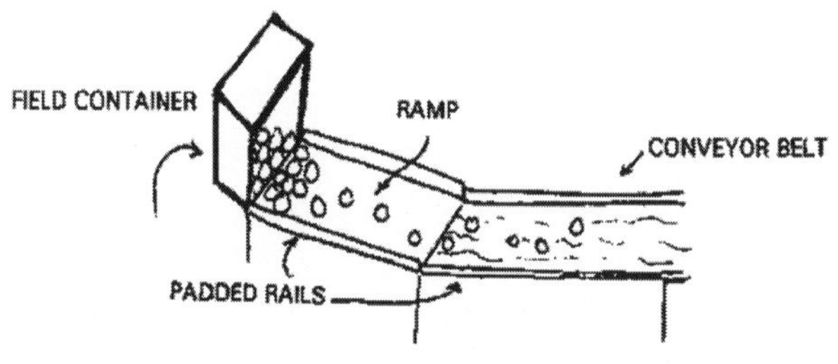

Figure 2. Dry dumping

Wet dumping is sometimes used to reduce mechanical damage, either by dumping into water rather than onto a dry ramp, or by immersion and floatation. If the specific density of the produce, such as apples, is lower than that of water the produce will float. For some produce, such as pears, salts must be added to the water to increase its specific density and assure fruit floatation.

The canvas curtain illustrated below is used to break the fall of fruit moving from a conveyor into a bulk bin.

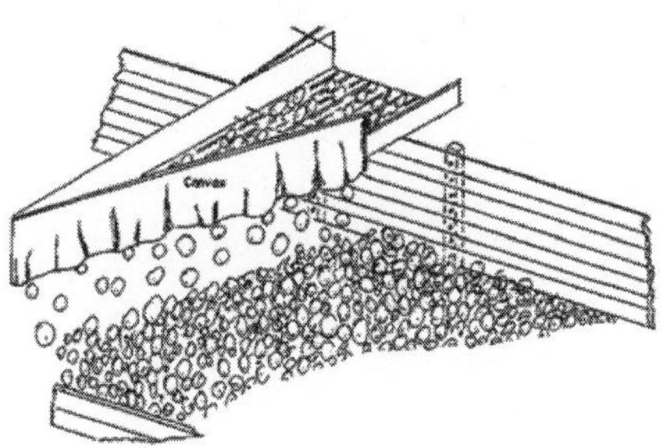

Figure 3. Canvas curtain

Washing

Steel drums can be used to make a simple washing stand. The drums are cut in halt fitted with drain holes and all the metal edges are covered with split rubber or plastic hose. The drums are then set into a sloped wooden table. The table top is constructed from wooden slats and is used as a drying rack before packing.Because steel drums are often used to store petroleum and chemical products, they should be thoroughly cleaned before being used as a washing stand.This tank for washing produce is made from galvanized sheet metal. A baffle made of perforated sheet metal is positioned near the drain pipe and helps to circulate water through the produce. Fresh water is added under pressure through a perforated pipe, which helps move floating produce toward the drain end of the tank for removal after cleaning.

Improvements to the design shown below might include a removable trash screen in front of the baffle, or a recirculating system for wash water.

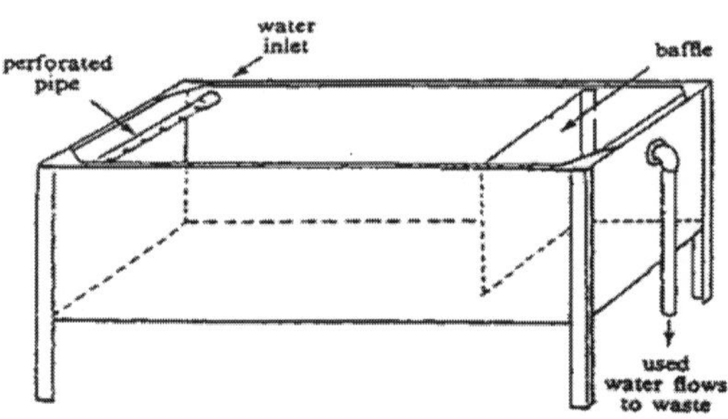

Figure 4.

WAXING

The waxing device illustrated here is designed to be used after a series of dry brushes on a conveyor line. Industrial wool felt is used to distribute the liquid wax to the fruits or vegetables from a trough made the same width as the belt. Evaporation of wax from the felt is reduced by covering the felt with a layer of heavy polyethylene sheeting.

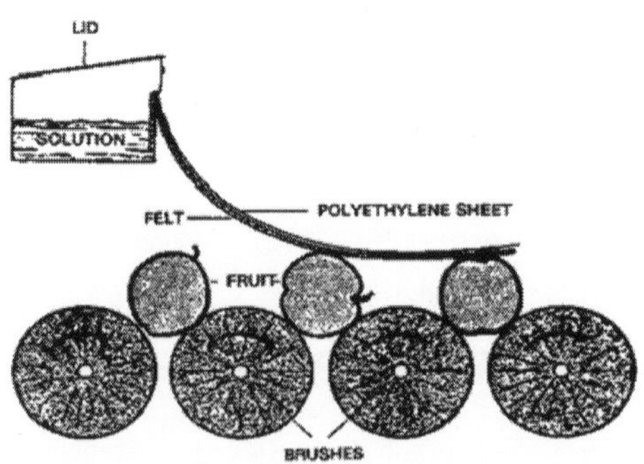

Figure 5.

SORTING/PACKING

Incoming produce is placed in the sorting bin, sorted by one worker into the packing bin, and finally packed by a second worker. If workers must stand to sort produce, a firm rubber pad for the floor can help reduce fatigue.

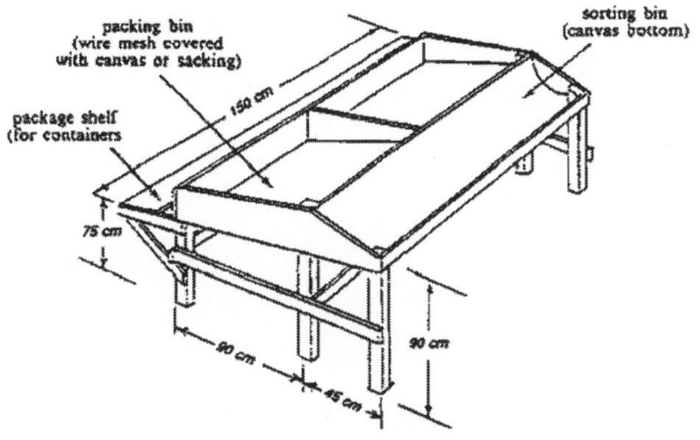

Figure 6.

The surface of the portable sorting table illustrated below is constructed from canvas and has a radius of about 1 meter (about 3 feet). The edges are lined with a thin layer of foam to protect produce from bruising during sorting, and the slope from the center toward the sorter is set at 10 degrees. Produce can be dumped onto the table from a harvesting container, then sorted by size, color and/or grade, and packed directly into shipping containers.

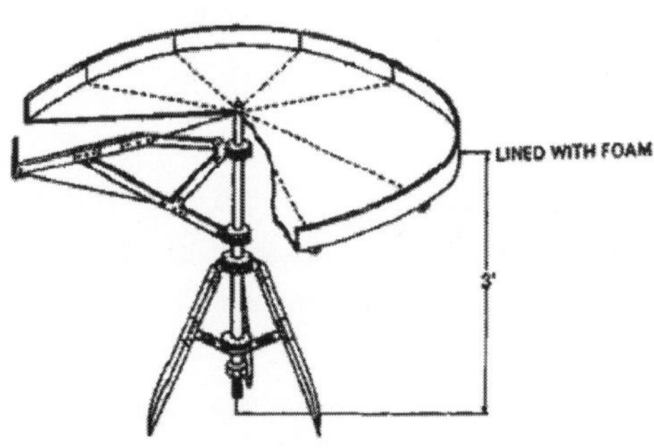

Figure 7.

Sorting

The following illustrations represent three types of conveyors used to aid sorting of produce. The simplest is a belt conveyor, where the sorter must handle the produce manually in order to see all sides and inspect for damage. A push-bar conveyor causes the produce to rotate forward as it is pushed past the sorters. A roller conveyor rotates the product backwards as it moves past the sorter.

Figure 8. Belt conveyor

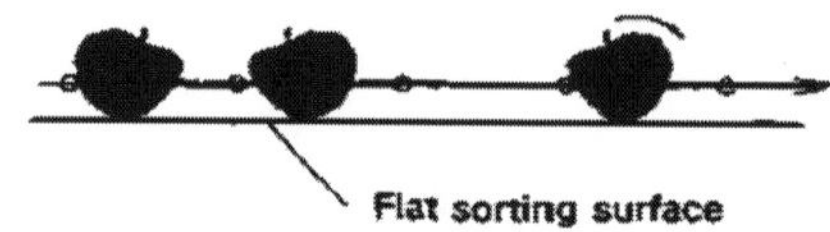

Figure 9. Push-bar conveyor

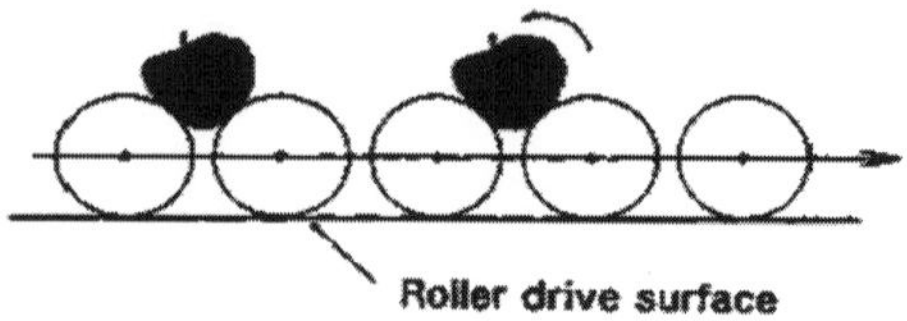

Figure 10. Roller conveyor

When sorting for rejects, and removing any product that is too small, decayed or damaged, the height of the sorting table should be set at a level comfortable for sorters. Stools, or a firm rubber pad on which to stand, can be provided to reduce fatigue. Locations of the table and the sorting bins should be chosen to minimize hand movements.

It is recommended that the workers' arms create a 45 degree angle when s/he reaches toward the table, and that the width of the table be less than 0.5 meter to reduce stretching. Good lighting will enhance the ability of the sorter to spot defects, and dark, dull belts or table tops can reduce eye strain.

If a conveyor system is in use, the product must not flow too fast for the sorters to do their work. The rotational speed of push-bar or roller conveyors should be adjusted to rotate the product twice within the immediate field of view of the worker.

Figure 11.

Sizing

Round produce units can be graded by using sizing rings. Rings can be fashioned from wood or purchased ready-made in a wide variety of sizes.

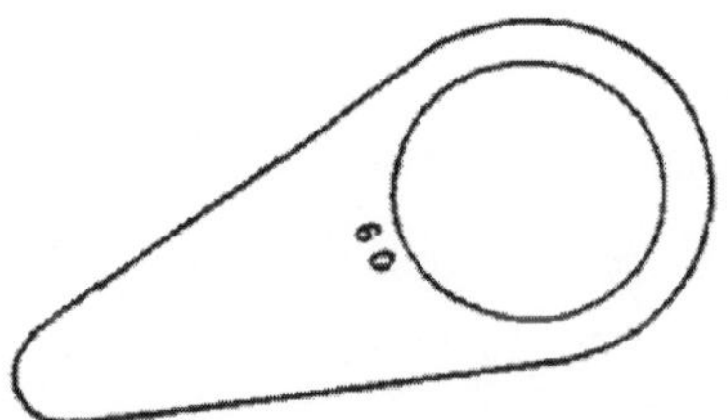

Figure 12. Single size hand held sizing ring

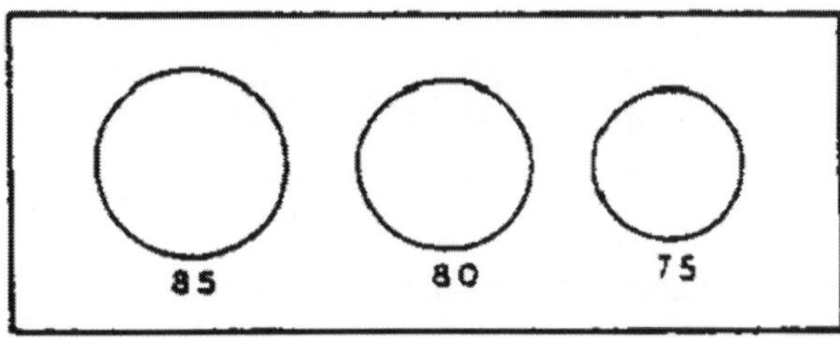

Figure 13. Multiple size rings

The rotary cylinder sizer is composed of five hollow cylinders which rotate in a counterclockwise motion when driven by an electric motor.

Each cylinder is perforated, with holes large enough to let fruits drop through. The first cylinder has the smallest diameter holes, and the fifth has the largest holes. When fruits fall through, they are caught on a slanted tray (the chute), and roll into the containers as shown. Take care that the distance of the drop is as short as possible to prevent bruising. Oversized fruits are accumulated at the end of the line. This equipment works best with round commodities.

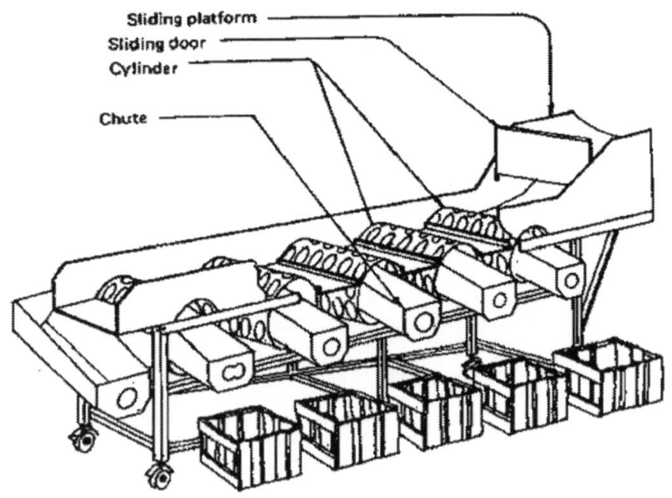

Figure 14. Rotary cylinder sizer

The onion sizing table illustrated below is one of three (or more) tables used in a stairway fashion. Each table is made of plywood, and has been perforated with holes of a specific size. The uppermost table has the largest size holes, and the lowest table has the smallest holes. A layer of onions is dumped onto the uppermost table. Those that do not pass through are classified as "extra-large'' in size. Those that pass through fall into a mesh bag and roll into a large container. This container of onions is dumped onto the second sizing table. The onions that do not pass through are classified as "large", and so on.

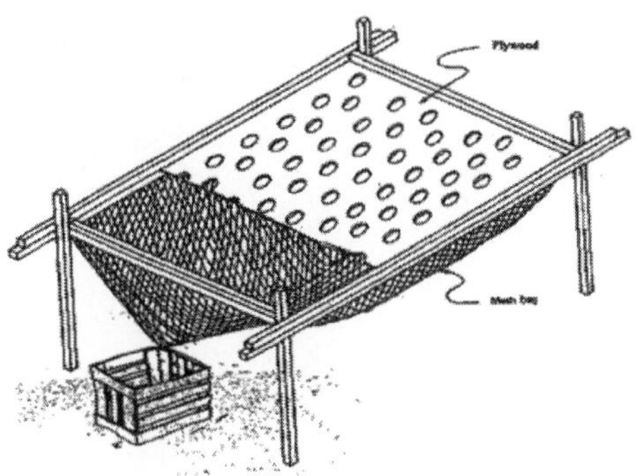

Figure 15. Onion sizing table

The pommelo sizer illustrated below is composed of a rectangular chute made of plywood, padded with foam to prevent bruising. The fruit is dumped into the octagonal platform at the top of the chute, then allowed to roll, one by one, down toward a series of constrictions. Large fruits are caught in the first constriction, medium in the second, and small in the last. Undersized fruit passes out the end of the chute directly into a container. Workers must manually remove each fruit and place it into the appropriate size container before the next fruit can pass through the chute. The sizing is fastest when five workers are stationed at the sizer.

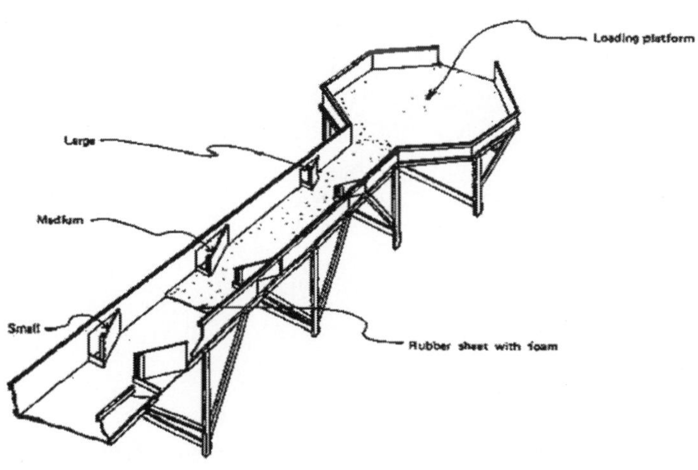

Figure 16. Pmmelo sizer

If a conveyor system is used in the packinghouse, a wide variety of sizing chains and belts are available for sorting produce. Sizing chains can be purchased in many widths and in any size opening.

Square openings are usually used for commodities such as apples, tomatoes and onions, while rectangular openings are used for peaches and peppers. Hexagonal openings are often used for potatoes and onions.

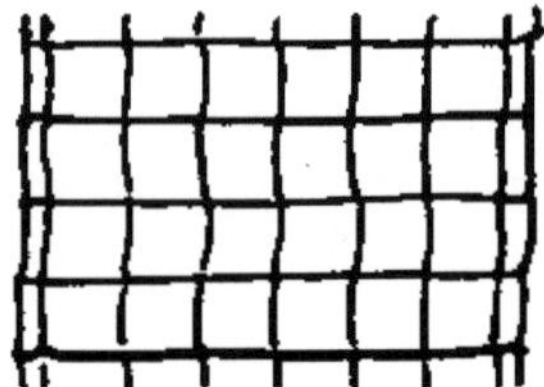

Figure 17. Square

Figure 18. Rectangular

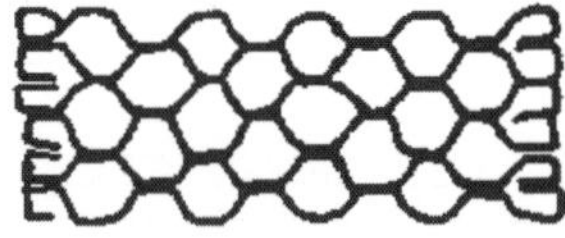

Figure 19. Hexagonal

Fruit Packing Line

Small scale equipment for packing produce is available from several manufacturers and suppliers. Illustrated below is a fruit packing line, available from TEW Manufacturing Corporation at a cost of less than US$ 5000. This particular model includes a receiving belt, washer and sorting table.

References

Smilanick, J.L and J. Usall i Rodie. 1995. Biological control of postharvest disseases of fresh fruit. *Perishables Handling Special Issue* No. 82: 19-20.

Thompson, A.K. 1996. *Postharvest Technology of Fruits and Vegetables*. Blackwell Science, Ltd., Oxford, UK, 410 pp.

Walker, D. J. 1992. *World Food Programme Food Storage Manual* Chatham, UK: Natural Resources Institute.

Wills, R., B. McGlasson, D. Graham, and D. Joyce. 1998. *Postharvest: An Introduction to the Physiology and Handling of Fruit, Vegetables, and Ornamentals.* CAB International, Wallingford, UK, 262 pp.

5

Packing and Packaging Materials

If produce is packed for ease of handling, heavily waxed cartons, wooden crates or rigid plastic containers are preferable to bags or open baskets, since bags and baskets provide no protection to the produce when stacked. Sometimes locally constructed containers can be strengthened or lined to provide added protection to produce. Waxed cartons, wooden crates and plastic containers, while more expensive, are reusable and can stand up to the high relative humidity found in the storage environment. Containers should not be filled either too loosely or too tightly for best results. Loose products may vibrate against others and cause bruising, while overpacking results in compression bruising. Shredded newspaper is an inexpensive and lightweight filler for shipping containers.

For small-scale handlers interested in constructing their own cartons from corrugated fibreboard, Broustead and New provide detailed information. Many types of agricultural fibres are suitable for paper making, and handlers may find it economically sensible to include these operations in their postharvest system.

Throughout the entire handling system, packaging can be both an aid and a hindrance to obtaining maximum storage life and quality. Packages need to be vented yet be sturdy enough to prevent collapse. Collapsed packages provide tattle or no protection, requiring the commodity inside to support all of the weight of the overhead load. Packing is meant to protect the commodity by immobilizing and cushioning it, but temperature management can be made more difficult if packing materials block ventilation holes. Packing materials can act as vapor barriers and can help maintain higher relative humidities within the package. In addition to protection, packaging allows quick handling throughout distribution and marketing and can minimize impacts of rough handling.

Produce can be hand-packed to create an attractive pack, often using a fixed count of uniformly sized units. Packaging materials such as trays, cups, wraps, liners and pads may be added to help immobilize the produce. Simple mechanical packing systems often use the volume-fill method or tight-fill method, in which sorted produce is delivered into boxes, then vibration settled. Most volume-fillers are designed to use weight as an estimate of volume, and final adjustments are done by hand.

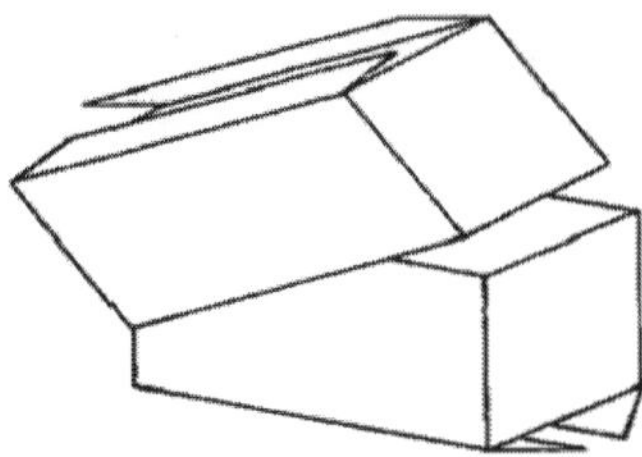

Packaging in plastic films can modify the atmosphere surrounding the produce (modified atmosphere packaging or MAP). MAP generally restricts air movement, allowing the product's normal respiration processes to reduce oxygen content and increase carbon dioxide content of the air inside the package. An additional major benefit to the use of plastic films is the reduction of water loss.

MAP can be used within a shipping container and within consumer units. Atmospheric modification can be actively generated by creating a slight vacuum in a vapor sealed package, and then replacing the package atmosphere with the desired gas mixture. In general, lowering oxygen and increasing carbon dioxide concentrations will be beneficial for most commodities. Selection of the best polymeric film for each commodity/ package size combination depends upon film permeability and the respiration rate of the commodity under the expected time/temperature conditions during handling. Absorbers of oxygen, carbon dioxide and/or ethylene can be used within packages to help maintain desired atmospheric composition.

Modified atmosphere packaging should always be considered as a supplement to proper temperature and relative humidity management. The differences between beneficial and harmful concentrations of oxygen and carbon dioxide for each kind of produce are relatively small, so great care must be taken when using these technologies.

Packing Practices

The packing stand can be bolted to a second stand of the same construction if more space is required for packing produce. When trimming is necessary, add a loose board, thick enough to reach above the height of the front rail. The front rail should be smooth and rounded.

A simple field packing station can be constructed from wooden poles and a sheet of polyethylene. Thatch over the roof will provide shade and keep the station cool. The structure should be oriented so that the roof overhang keeps out the majority of the sun's rays.

Hands of bananas, after undergoing washing to remove latex and perhaps spraying with fungicides, are typically packed into cardboard containers lined with polyethylene. The following illustrations depict one method of filling a container with the fruit in order to ensure less damage during transport. Note that the polyethylene liner is folded up over the bananas before closing the box.

Wide, flat medium-to-small hand in middle of compartment

Medium-length, wide hand on top, crown not touching fruit below

Medium-to-short length, wide hand, crown not touching fruit below

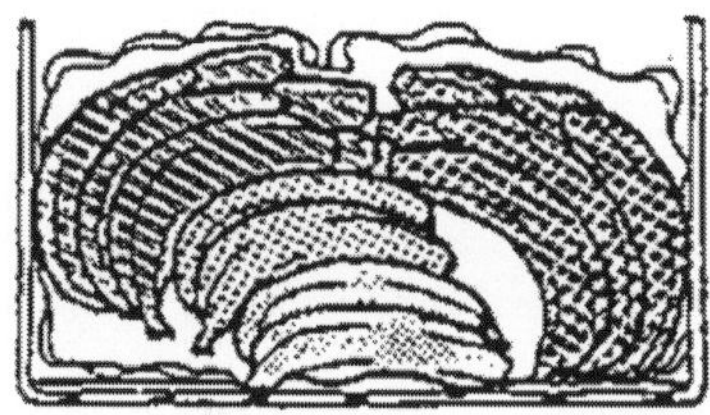

One large hand, or two clusters with long fingers

A circular rotating table can be used to pack a variety of crops. The produce is fed in along a conveyor or if no conveyor is in use, simply put onto the table, where packers select the produce and fill cartons at their stations. In the illustration below, a discard belt has been added below the supply belt, allowing easy disposal of culls.

Each packer can work independently, trimming as needed and check weighing cartons on occasion.

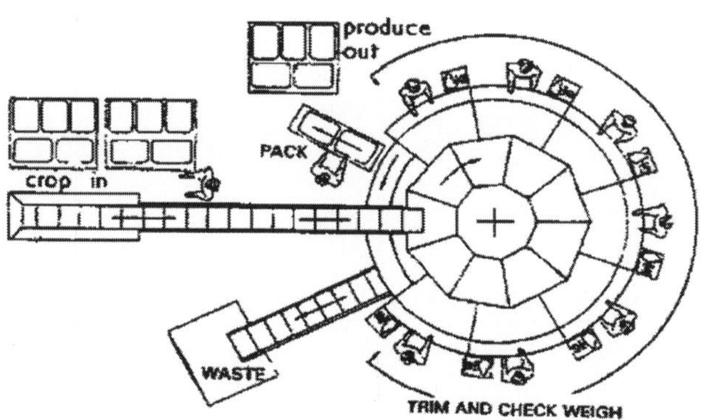

Packing Containers

There are many types of packing containers. The three containers illustrated below are constructed from corrugated cardboard. The regular slotted container is fully collapsible and the most economical.

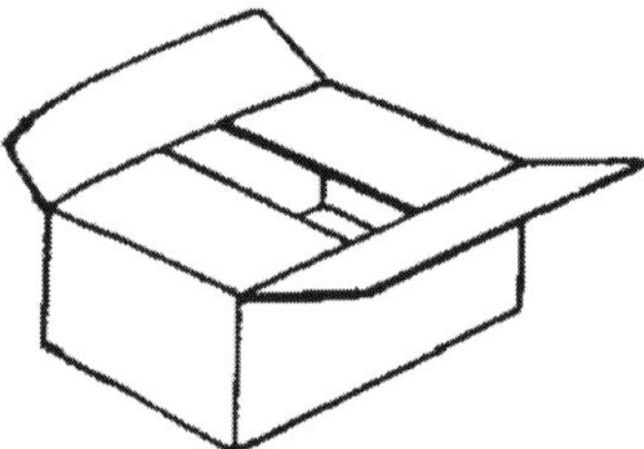

Telescopic containers (half or full) have the highest stacking strength and protect against bulging but are more costly.

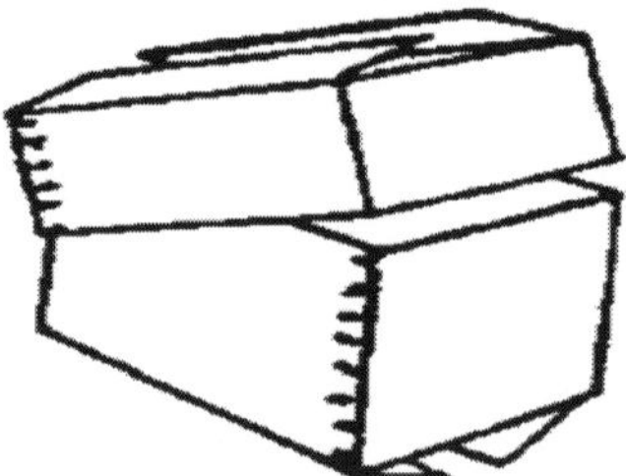

The container known as a Bliss box has very strong corners, but is not collapsible.

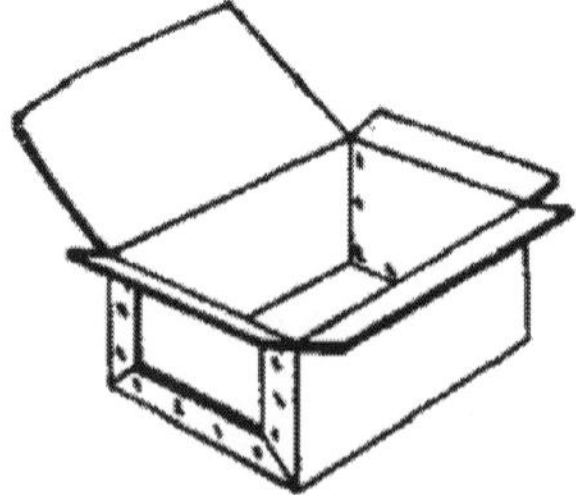

Sacks are often used to package produce, since they tend to be inexpensive and readily available.

The following table provides some information regarding the characteristics of different kinds of materials used to make sacks.

Sack types	*Tearing and snagging*	*Impact*	*Protection against*		*Contamination*	*Notes*
			Moisture absorption	*Insect invasion*		
Jute	Good	Good	None	None	Poor, also cause contamination by sack fibres	Bio-deterioration Insect harbourage. Odour retention.
Cotton	Fair	Fair	None	None	Fair	High re-use value.
Woven plastics	Fair-Good	Good	None	Some protection	Fair	Badly affected by ultra-violet light. Difficult to stitch.
Paper	Poor	Fair-Poor	Good - WFP multiwall sacks have plastic liners.	Some protection , better if treated.	Good	Consistent quality. Good print.

Paper or cloth sacks can be easily closed using a length of strong wire and twisting tool.

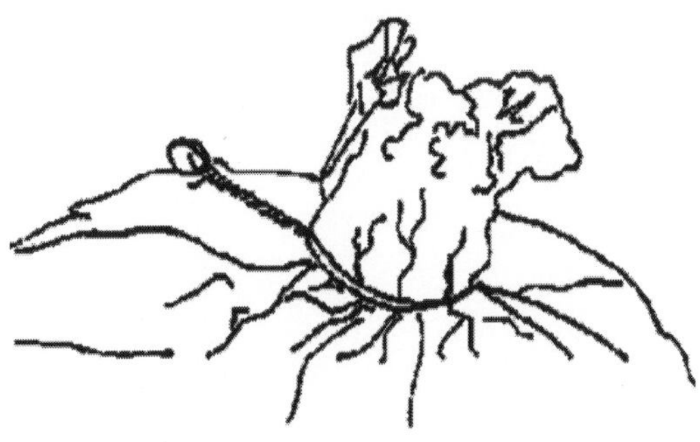

The table below provides examples of some types of typical mechanical damage and their effect on packaging containers.

Type of damage	*Container*	*Result*	*Important factors*
Impact damage through dropping	Sacks - woven and paper	Splitting of seams and material causing leaking and spillage loss.	Seam strength
	Fibreboard boxes	Splitting of seams, opening of flaps causing loss of containment function. Distortion of shape reducing stacking ability.	Bursting strength Closure method
	Wooden cases	Fracture of joints, loss of containment function.	Fastenings Wood toughness
	Cans and drums	Denting, rim damage. Splitting of seams and closures causing loss of containment and spoilage of contents.	
	Plastic bottles	Splitting or shattering causes loss of contents.	Material grade Wall thickness
Compression damage through high stacking	Fibreboard boxes	Distortion of shape, seam splitting causing loss of containment and splitting of inner cartons, bags, and foil wrappings.	Box compression strength
	Plastic bottles	Distortion, collapse and sometimes splitting, causing loss of contents.	Design, material, wall thickness
Vibration	Woven sacks	Sifting out of contents.	Closeness of weave
	Corrugated fibreboard cases	Become compressed and lose their cushioning qualities. Contents more prone to impact damage.	Box compression strength
Snagging, tearing, hook damage	Sacks - woven and paper	Loss of containment function - spillage (more severe with paper sacks).	Tear strength
	Tins	Punctured, loss of contents.	Metal thickness

The diagrams below are for a variety of fibreboard containers.

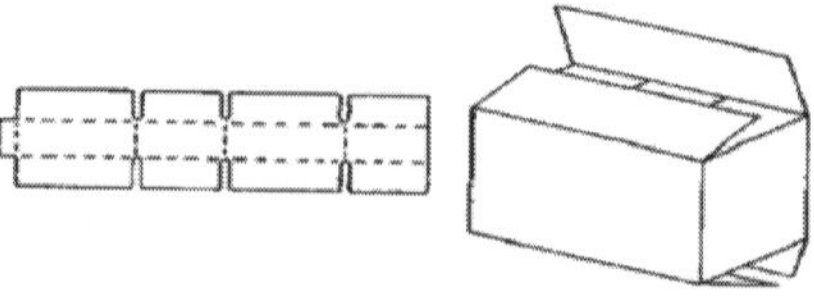

One piece box

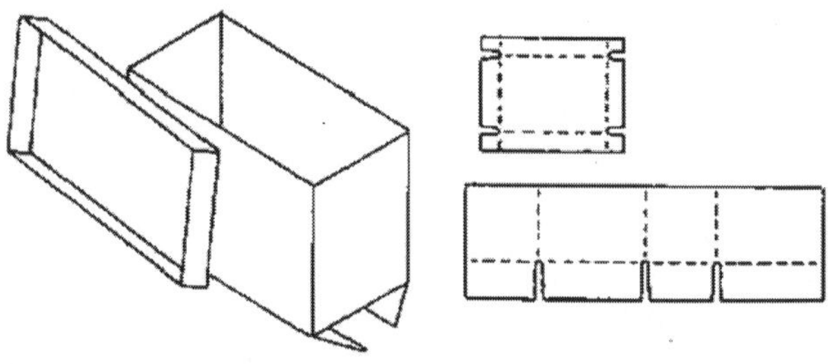

Two-piece box with cover

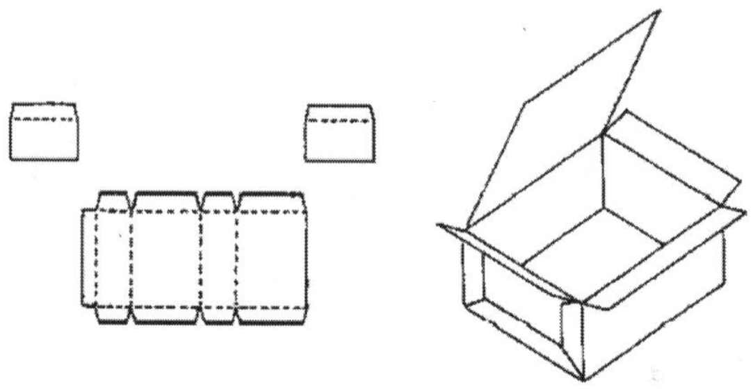

Bliss-style box

The diagrams below are for a variety of commonly used fibreboard containers. Final dimensions can be altered to suit the needs of the handler.

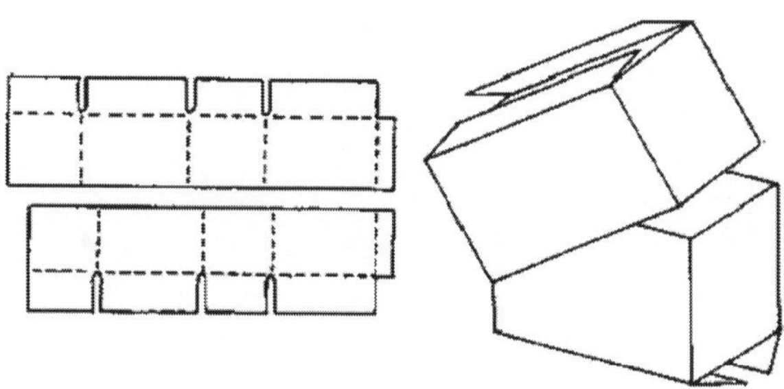

Full telescoping box

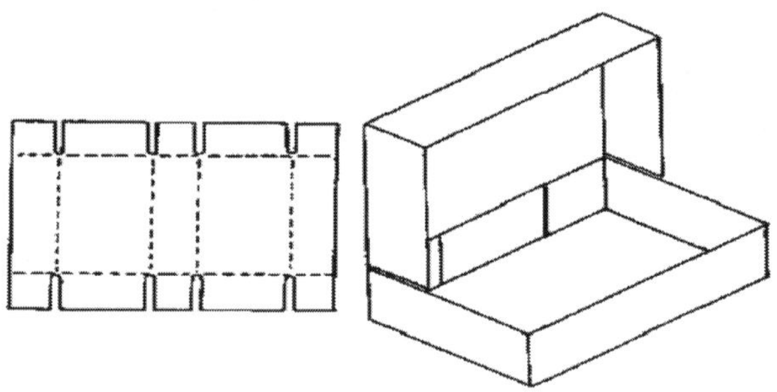

One-piece telescoping box

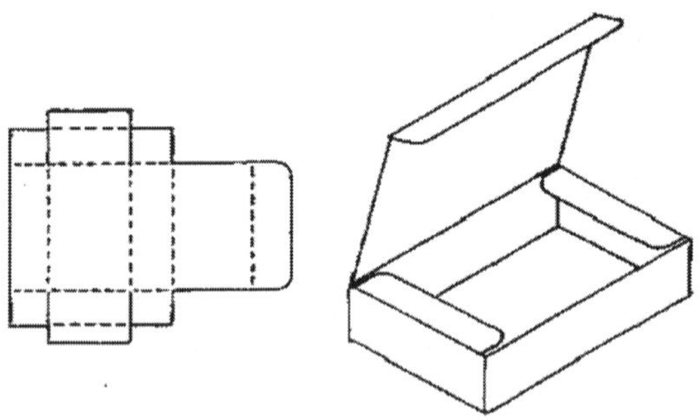

One-piece tuck-in cover box

The diagrams below are for a variety of commonly used fibreboard containers. Final dimensions can be altered to suit the needs of the handler.

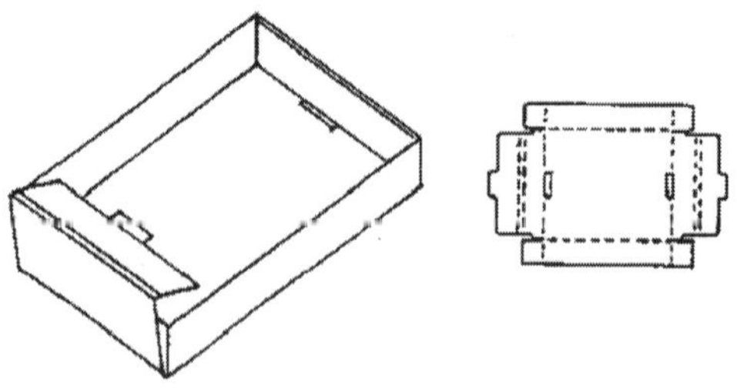

Self-locking tray

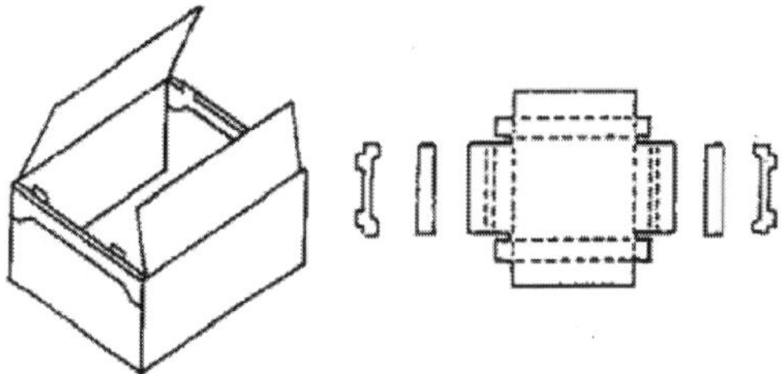

Interlocking box

Shipping containers can be designed and made by the user from fibreboard in any size and shape desired. Three types of joints are commonly used to construct sturdy boxes.

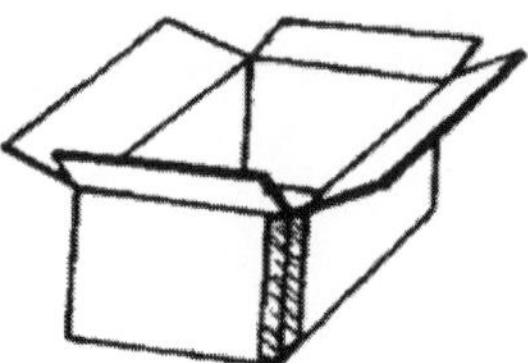

Taped joints

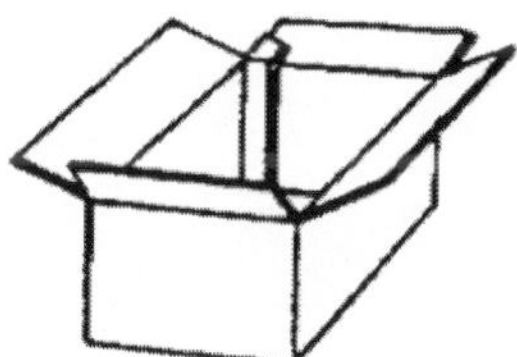

Glued Joints

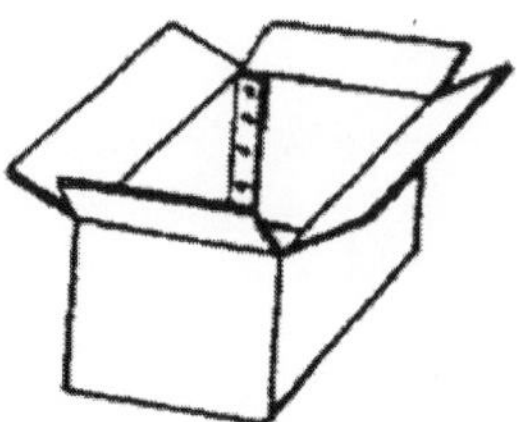

Stapled joints

Containers can be constructed from wood and wire, using the general diagrams provided below. A special closing tool makes bending the wire loops on the crate's lid easier for packers to do. Wirebound crates are used for many commodities including melons, beans, eggplant, greens, peppers, squash and citrus fruits. Package Research Laboratory can provide a list of suppliers in your area.

A wooden lug is the typical packing container for table grapes. This container is very sturdy and maintains its stacking strength over long periods of time at high relative humidity. Rigid plastic containers are also widely used.

Often, a paper liner is folded over the grapes before the top is nailed closed. The liner protects the produce from dust and water condensation. If a pad containing sulfur dioxide can be enclosed with the grapes within a plastic liner as a treatment to control decay. Most commodities other than table grapes can be damaged (bleached) by sulfur dioxide treatments.

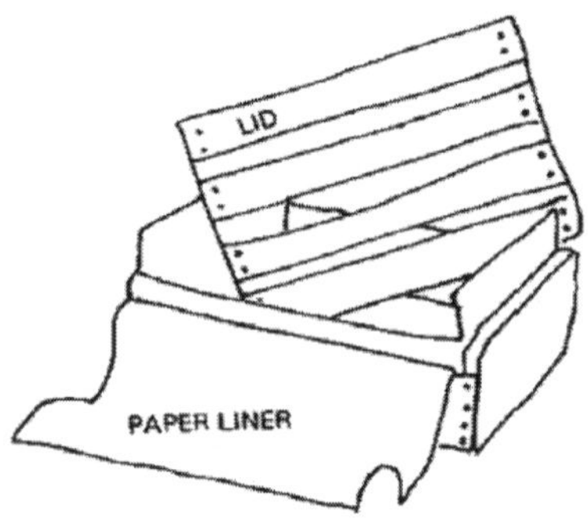

Rigid plastic or wooden containers are also used extensively for asparagus. The trimmed spears are packed upright in containers that provide for a large amount of ventilation.

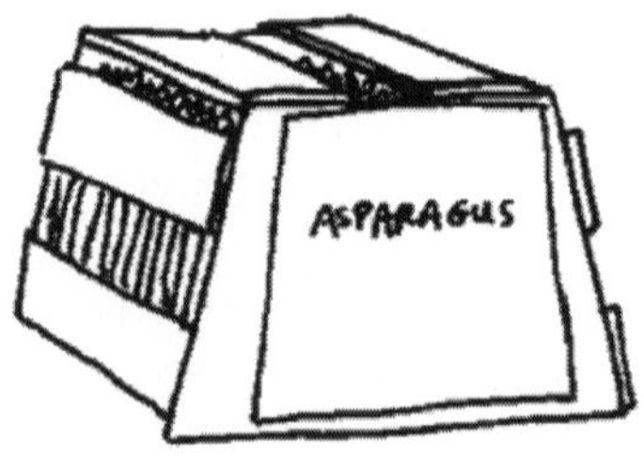

Containers for cut flowers are often long and narrow, of full telescopic design with vents at both ends to facilitate forced-air cooling. The total vent area should be 5% of the total box surface area. A closable flap can help maintain cool temperatures if boxes are temporarily delayed in transport or storage in an uncontrolled temperature environment.

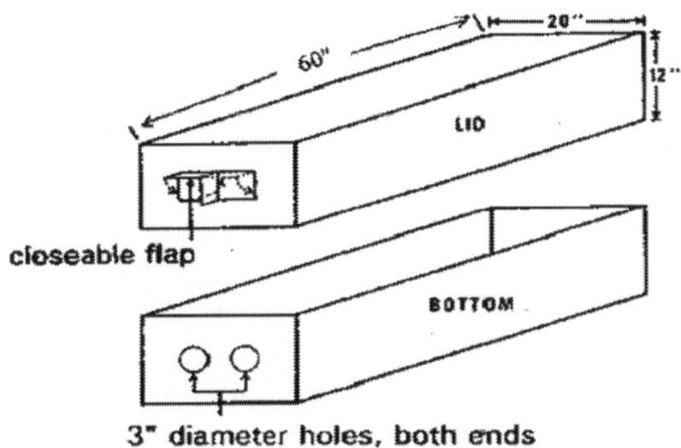

A simple wooden tray with raised corners is stackable and allows plenty of ventilation for fragile crops such as ripe tomatoes.

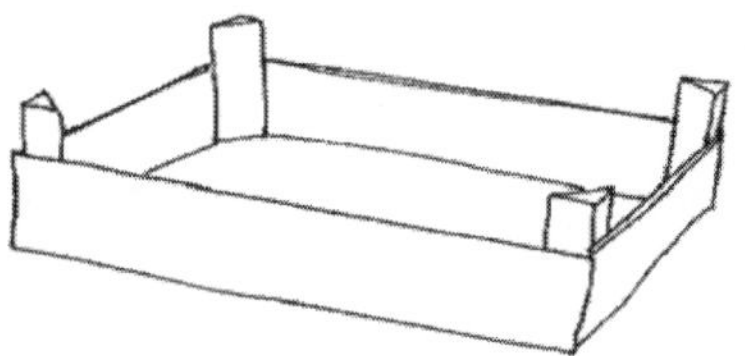

Packaging practices

Adding a fiberboard divider to a carton will increase stacking strength. The use of dividers is common with heavy crops such as melons. The dividers also prevent melons from vibrating against one another during handling and transport. Wooden inserts, or fiberboard folded into triangles and placed in all four corners can be especially useful when a carton needs strengthening.

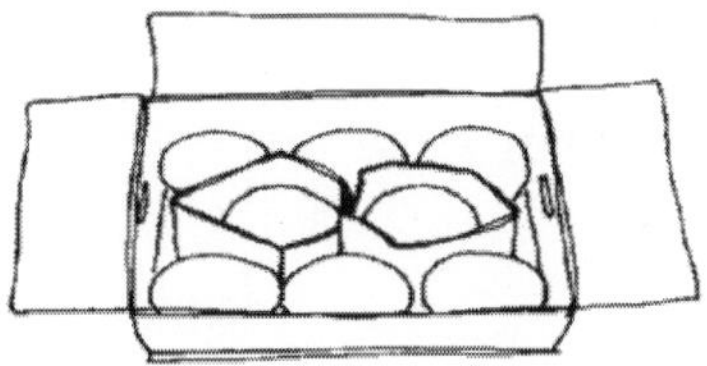

Fiberboard divider

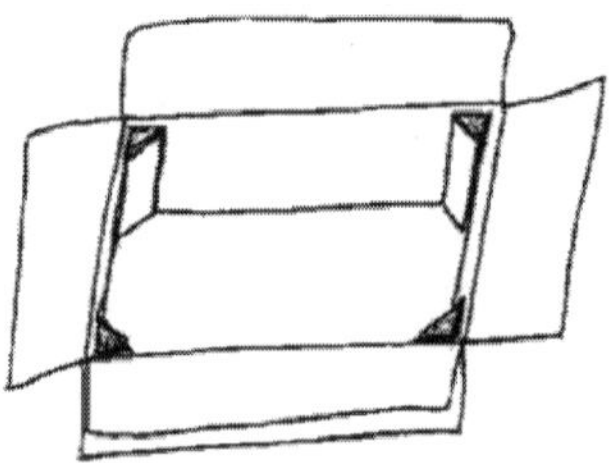

Triangular corner supports

When locally made containers have sharp edges or rough inner surfaces, a simple, inexpensive inner made from fiberboard can be used to protect produce from damage during handling.

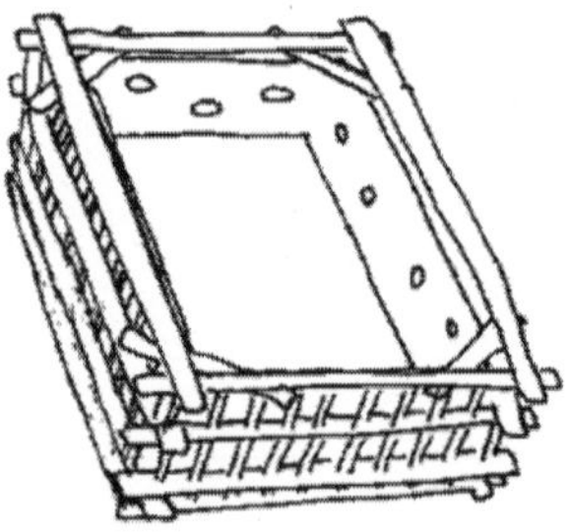

Cardboard liner for a palm rib crate

If large bags or baskets must be used for bulk packaging of fruits or vegetables, the use of a simple vent can help reduce the buildup of heat as the product respires. In the illustration below, a tube of woven bamboo (about one meter long) is used to vent a large bag of chili peppers.

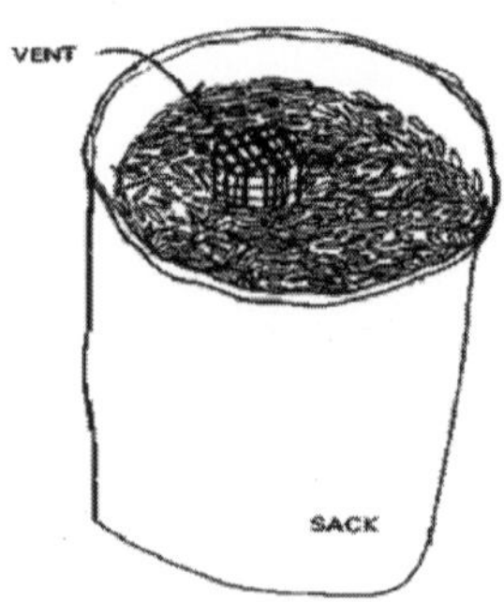

Thin paper or plastic sleeves are a useful material for protecting cut flowers from damage during handling and transport. In the illustration below, a packer is pulling a sleeve up over a bunch of flowers before packing the flowers into a vented fibreboard carton. The sleeves both provide protection and help keep the bunches of flowers separate inside the box.

LABELLING

Labeling packages helps handlers to keep track of the produce as it moves through the postharvest system, and assists wholesalers and retailers in using proper practices. Labels can be preprinted on fiberboard boxes, or glued, stamped or stenciled on to containers. Brand labeling packages can aid in advertising for the product's producer, packer and/or shippers. Some shippers also provide brochures detailing storage methods or recipes for consumers.

Shipping labels can contain some or all of the following information:

— Common name of the product.

— Net weight, count and/or volume.

— Brand name.

— Name and address of packer or shipper.

— Country or region of origin.
— Size and grade.
— Recommended storage temperature.
— Special handling instructions.
— Names of approved waxes and/or pesticides used on the product.

Labeling of consumer packages is mandatory under FDA regulations. Labels must contain the name of the product, net weight, and name and address of the producer, packer or distributor.

Modularization of Containers

When a variety of different sized cartons are packed at the same time, using boxes in standard sizes can greatly ease future handling. When handling boxes that are non-uniform, stacks can be unstable or heavier cartons can crush lighter ones. An unstable load is likely to fall over during transport or to collapse during storage.

Recommended container sizes are shown below. These containers are part of the MUM program (Modularization, Unitization and Metrication) advocated by the USDA. They can all be stacked in a variety of patterns, depending upon their size, yet still form a stable load on a single pallet of 1000 x 1200 mm (40 x 48 inches).

MUM containers for horticultural crops:

Outside Dimensions		*Number Per Layer*	*Pallet Surface Area Utilized*
mm	*Inches*		*Percentage*
600 x 500	(23.62 x 19.69)	4	100
500 x 400	(19.68 x 15.75)	6	100
600 x 400	(23.62 x 15.75)	5	100
500 x 333	(19.68 x 13.11)	7	97
600 x 333	(23.62 x 13.11)	6	99
500 x 300	(19.68 x 11.81)	8	100
475 x 250	(18.70 x 9.84)	10	99
400 x 300	(15.75 x 11.81)	10	100
433 x 333	(17.01 x 13.11)	8	96
400 x 250	(15.74 x 9.84)	12	100

An example of a pallet load of MUM container

Modified atmosphere packaging (MAP)

Within a consumer package: If commodity and film permeability characteristics are properly matched, an appropriate atmosphere can evolve passively through consumption of O_2 and production of CO_2 during respiration. Some rigid plastic consumer packages are designed with a gas diffusion window.

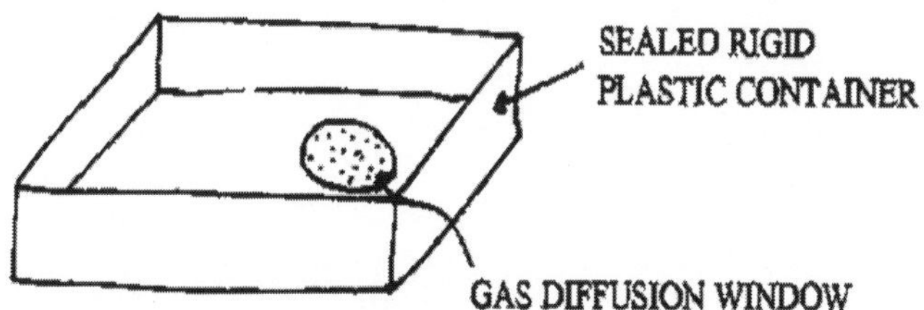

Lightly processed lettuce (shredded or chopped) can be packaged in 5-mil plastic bags. After a partial vacuum is created, a gas mixture of 30 to 50% O_2 and 4 to 6% CO_2 is introduced into the bag, which is then sealed.

Sealed Plastic Bag

Within a shipping container: Polyethylene liners are added to shipping containers in cherry boxes, and polyethylene bags are used for bananas destined for distant markets.

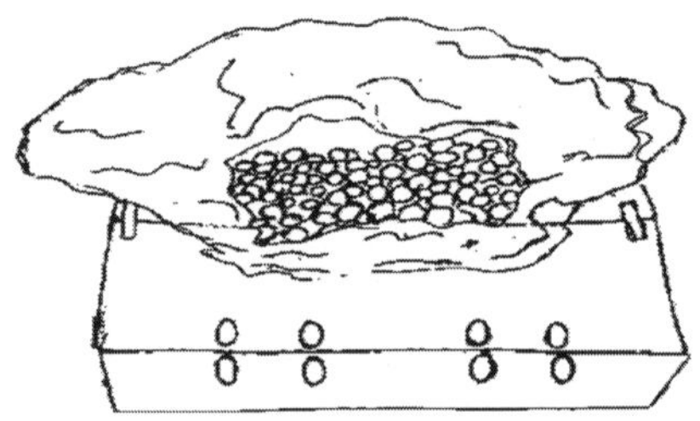

Plastic Liner

Within a pallet: A single pallet load of produce such as strawberries can be sealed within a shroud of 5 mil polyethylene bag and a plastic sheet on the pallet base using wide tape. A slight vacuum can be introduced and 15% CO_2 added to the air introduced via a small hose.

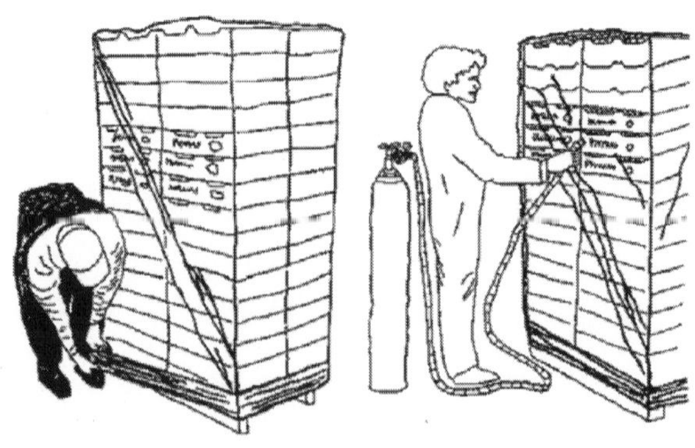

Many plastic films are available for packaging, but very few have gas permeabilities that make them suitable for MAP. Low density polyethylene and polyvinyl chloride are the main films used in packaging fresh fruits and vegetables. Saran and polyester have such low gas permeabilities that they are suitable only for commodities with very low respiration rates.

The following table provides the permeabilities of the films currently available for packaging fresh produce.

Film type	*Permeabilities (cc/m2/mil/day at 1 atm)*		*CO_2:O_2 Ratio*
	CO_2	O_2	
Polyethylene: low density	7,700-77,000	3,900-13,000	2.0-5.9
Polyvinyl chloride	4,263-8,138	620-2,248	3.6-6.9
Polypropylene	7,700-21,000	1,300-6,400	3.3-5.9
Polystrene	10,000-26,000	2,600-7,700	3.4-3.8
Saran	52-150	8-26	5.8-6.5
Polyester	180-390	52-130	3.0-3.5

Unit loads

Many shippers and receivers prefer to handle unit loads of produce pallets rather than handling individual shipping containers. The switch to unit loads has reduced handling, causes less damage to the containers and produce inside, and allows faster loading/ unloading of transportation vehicles.

If small scale handlers wish to use unit loads for shipping produce, either wooden pallets or slip sheets can serve as the base of the load. Using guides for aligning the boxes (such as placing the pallet to be loaded against the corner of a room, or building a set of "bounce boards" if the pallet is loaded outside) will stabilize the load.

Using fiberboard, plastic or wooden containers with vertical interlocking tabs can also help improve the unit load's stability. Containers must have holes for ventilation which align when stacked squarely on top of one another. Glue can be used between layers of containers to reduce slipping, and plastic netting or plastic or metal straps should be added to secure the load. Cornerboards made from fiberboard, plastic or metal help to provide for a stable unit load.

References

Grierson, W. 1987. *Postharvest Handling Manual, Commercialization of Alternative Crops Project*. Belize Agribusiness Company/USAID/Chemonics International Consulting Division, 2000 M Street, N.W., Suite 200, Washington, D.C. 20036.

Kader, AA (ed). 2002. *Postharvest Technology of Horticultural Crops* (3rd Edition). UC Publication 3311. University of California, Division of Agriculture and Natural Resources, Oakland, California 94608. 535 pp.

Moline, HE. 1984. *Postharvest Pathology of Fruits and Vegetables: Postharvest Losses in Perishable Crops.* U.C. Bulletin 1914, University of California, Division of Agriculture and Natural Resources, Oakland, California 94608.

Thompson, J.F. et al. 1998. *Commercial cooling of fruits, vegetables, and flowers*. Publ. 21567, ANR, University of California, Oakland, CA, 61 pp.

6

Decay and Pest Control

The first line of defense against insects and disease is good management during production. The second is careful harvesting and preparation for market in the field. Thirdly, sorting out damaged or decaying produce will limit contamination of the remaining, healthy produce. Yet, even when the greatest care is taken, sometimes produce must be treated to control insects or decay-causing organisms.

Certain fungi and bacteria in their germination phase are susceptible to cold, and infections can be reduced by treating produce with a few days of storage at the coldest temperature the commodity can withstand without incurring damage. Rhizopus stolonifer and Aspergillus niger (black mold) can be killed when germinating by 2 or more days at 0°C (32 F). On the other hand, brief hot water dips or forced-air heating can also be effective, especially for reducing the microbial load for crops such as plums, peaches, papaya, cantaloupe and stone fruits, sweetpotatoes and tomatoes.

While high humidity in the storage environment is important for maintenance of high quality produce, free water on the surface of commodities can enhance germination and penetration by pathogens. When cold commodities are removed from storage and left at higher ambient temperatures, moisture from the surrounding warm air condenses on the colder product's surfaces. A temporary increase in ventilation rate (using a fan) or increasing exposure of the commodity to drier air can help to evaporate the condensed moisture and to reduce the chances of infection.

Cold treatments can also serve to control some insect pests, and are currently used for the control of fruit flies. Treatment requires 10 days at 0

C (32 F) or below, or 14 days at 1.7 C (35 F) or below, so treatment is only suited to commodities capable of withstanding long-term low-temperature storage such as apples, pears, grapes, kiwifruit and persimmons. For produce packed before cold storage treatment, package vents should be screened to prevent the spread of insects during handling.

Hot water dips or heated air can be used for direct control of postharvest insects. In mangoes, an effective treatment is 46.4 C for 65 to 90 minutes, depending on size. Fruit should not be handled immediately after heat treatment. Whenever heat is used with fresh produce, cool water showers or forced cold air should be provided to help return the fruits to their optimum temperature as soon as possible after completion of the treatment.

Control of storage insects in nuts and dried fruits and vegetables can be achieved by freezing, cold storage (less than 5 C or 41 F)), heat treatments, or the exclusion of oxygen (0.5% or lower) using nitrogen. Packaging in insect-proof containers is needed to prevent subsequent insect infestation.

Some plant materials are useful as natural pesticides. Cassava leaves are known to protect harvested cassava roots from pests when used as packing material in boxes or bags during transport and short-term storage. It is thought that the leaves release cyanogens, which are toxic to insects. The ashes of the leaves of Lantana spp. andOchroma logopur have been found to be very effective when used as a dust against aphids attacking stored potatoes. The pesticidal properties of the seeds of the neem tree (as an oil or aqueous extract) are becoming more widely known and used throughout the world. Native to India, neem acts as a powerful pesticide on food crops but appears to be completely non-toxic to humans, mammals and beneficial insects Any "natural pesticide" must be shown to be safe for humans before its approval by regulatory authorities.

Chemical Controls

Washing produce with chlorinated water can prevent decay caused by bacteria, mold and yeasts on the surface of produce. Calcium hypochlorite (powder) and sodium hypochlorite (liquid) are inexpensive and widely available. The effectiveness of the treatment will be decreased if organic matter is allowed to build up in the wash water. The effectiveness of chlorine increases as pH is reduced from pH 11 to pH 8, but at lower pH chlorine becomes unstable.

Fruits and vegetables can be washed with hypochlorite solution (25 ppm available chlorine for two minutes) then rinsed to control bacterial decay. Alternatively, these commodities can be dipped in hypochlorite solution (50 to 70 ppm available chlorine) then rinsed with tap water for control of bacteria, yeasts and molds. There are some chemicals that are generally recognized safe (GRAS) which are used to control a variety of molds and fungi on fruit crops.

Sulfur

Sulfur is used on bananas as a paste (0.1 % active ingredient) to control crown rot fungi.

Sulfur dioxide (SO_2) is used as a fumigant or a water spray (0.5% for 20 minutes for the initial treatment, then 0.2% for 20 minutes at 7 day intervals) on grapes to control Botrytis, Rhizopus and Aspergillus fungi.

Careful calculation of the amount of sulfur dioxide required to treat grapes can greatly reduce the need to vent or scrub the storage air after fumigation to remove excess SO_2.

Sodium or Potassium Bisulfite

Bisulfites are used in a sawdust mixture (usually contained within a pad that can be placed inside a carton) to release SO_2 for control of molds on grapes.

Bacterial soft rot of cabbage can be controlled by using lime powder or a 15% solution of alum (aluminum potassium sulfate) in water. After treatment of the butt-end of the cabbage heads, the produce should be allowed to dry for 20 to 30 minutes before packing.

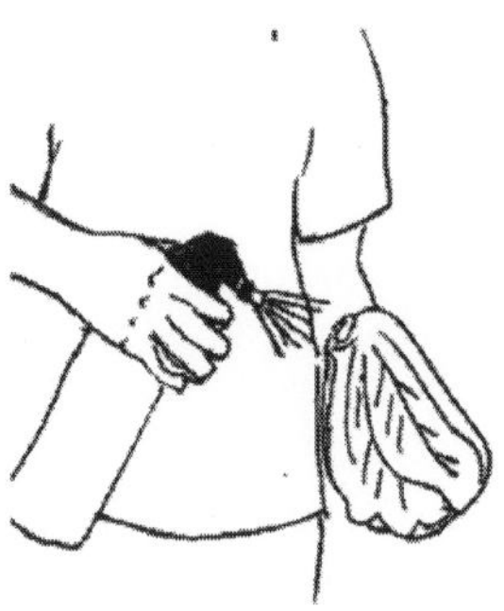

Applying alum solution (spray or brush on

On occasions when fungicides must be applied to produce, a simple tray with holes punched in the bottom can be used hold the commodity while it is sprayed.

When fruit is packed for export, fungicides are often applied to meet the requirements of international quality standards and to reduce deterioration during transport.

Fruit in a perforated plastic tray is introduced on a roller conveyor belt into the applicator. Inside a simple fan shaped deflector creates a curtain of liquid fungicide. The fruit passes under the curtain where it is drenched, then out of the applicator to drain on a tilted return tray. The tank holds up to 50 liters of fungicide solution, and a pump is mounted at the level of the tank outlet. A filter is fitted on the top of the tank to remove foreign matter from the return flow of fungicide from the applicator box and the return tray.

Controlled/Modified Atmosphere Treatments

For commodities that tolerate high CO_2 levels, 15 to 20% CO_2-enriched air can be used as a fungistat to control decay-causing pathogens, such as Botrytis cinerea on strawberry, blueberry, blackberry, fresh fig and table grapes during transport. Insecticidal atmospheres have been shown to be an effective substitute for methyl bromide fumigation to disinfest dried fruits, nuts and vegetables The effectiveness of insecticidal atmospheres depends upon the temperature, relative humidity, duration of exposure and life stage of the insect. Following are some examples:

1) *Sweetpotato weevil* (Cylas formicarius elegantulus) has been controlled at ambient temperature in stored tropical sweetpotatoes by treatment with low oxygen and high carbon dioxide atmospheres. At 25 C (76 F), storage in 2 to 4% oxygen and 40 to 60% carbon dioxide results in mortality of adult weevils in 2 to 7 days.
2) *Codling moth* (Cydia pomonella) in stone fruits can be controlled at 25 C (76 F) by using atmospheres of 0.5% oxygen and 10% carbon dioxide for 2 to 3 days or 6 to 12 days. Normal color and firmness changes during ripening are not affected by treatment.

Heat Treatments

Postharvest heating using hot water or hot forced-air to kill or weaken pathogens can be used as a method for decay control in fresh fruits and vegetables.

Table 1. Heat Treatments

Commodity	*Pathogens*	*Temperature (C)*	*Time (min)*	*Possible injuries*
Apple	*Gloeosporium* sp.	45	10	Reduced storage life
	Penicillium expansum			
Grapefruit	*Phytophthora citrophthora*	48	3	
Green beans	*Pythium butleri*	52	0.5	
	Sclerotinia sclerotiorum			
Lemon	*Penicillium digitatum*	52	5-10	
	Phytophthora sp.			
Mango	*C. gloeosporioides*	52	5	No stem rot control
Melon	Fungi	57-63	0.5	
Orange	*Diplodia* sp.	53	5	Poor degreening
	Phomopsis sp.			
	Phytophthora sp.			
Papaya	Fungi	48	20	
Peach	*Monolinia fructicola*	52	2.5	Motile skin
	Rhizopus stolonifer			
Pepper (bell)	*Erwinia* sp.	53	1.5	Slight spotting

Table 2. Hot Forced-Air Treatments

Commodity	*Pathogens*	*Temperature (C)*	*Time (min)*	*RH(%)*	*Possible injuries*
Apple	*Gloeosporium* sp.	45	15	100	Deterioration
	Penicillium expansum				
Melon	Fungi	30-60	35	low	Marked breakdown
Peach	*Monolinia fructicola*	54	15	80	
	Rhizopus stolonifer				
Strawberry	*Alternaria* sp.	43	30	98	
	Botrytis sp., *Rhizopus* sp.				
	Cladosporium sp.				

References

Mitra, S., ed. 1997. *Post-harvest physiology and storage of tropical and subtropical fruits.* Wallingford, UK: CAB International, 423 pp.

Shewfelt R.L. and Prussia, S.E. (eds). 1993. *Postharvest Handling A Systems Approach.* San Diego: Academic Press, 358 pp.

Smilanick, J.L 1995. *Status of postharvest fungicides and growth regulators.* Perishables Handling Special Issue No. 82: 30-32.

Throughout the period between harvest and consumption, temperature control has been found to be the most important factor in maintaining product quality. Fruits, vegetables and cut flowers are living, respiring tissues separated from the parent plant. Keeping products at their lowest safe temperature will increase storage life by lowering respiration rate, decreasing sensitivity to ethylene and reducing water loss. Reducing the rate of water loss slows the rate of shriveling and wilting, causes of serious postharvest losses. It is important to avoid chilling injury, since symptoms include failure to ripen (bananas and tomatoes), development of pits or sunken areas (oranges, melons and cucumbers), brown discoloration (avocados, cherimoyas, eggplant), increased susceptibility to decay (cucumbers and beans), and development of off-flavors (tomatoes).

If a ready supply of electricity is available, mechanical refrigeration systems provide the most reliable source of cold. Methods include room cooling, forced-air cooling and evaporative cooling. A variety of portable forced-air coolers have been designed for use by small-scale growers and handlers. However, a variety of simple methods exist for cooling produce where electricity is unavailable or too expensive. Some examples of alternative systems include night air ventilation, radiant cooling, evaporative cooling, the use of ice and underground (root cellars, field clamps, caves) or high altitude storage.

Cooling involves heat transfer from produce to a cooling medium such as a source of refrigeration. Heat transfer processes include conduction, convection, radiation and evaporation.

Several simple practices are useful for cooling and enhancing storage system efficiency wherever they are used, and especially in developing countries, where energy savings may be critical. Shade should be provided over harvested produce, packing areas, for buildings used for cooling and storage and for transport vehicles. Using shade wherever possible will help to reduce the temperatures of incoming produce. Trees are a fine source of shade and can reduce ambient temperatures around packinghouses and storage areas. Light colors on buildings will reflect light (and heat) and reduce heat load. Sometimes spending money will save money, as when purchasing lighting equipment. High pressure sodium lights produce less heat and use less energy than incandescent bulbs.

Another aspect to consider when handling fruits and vegetables is the relative humidity of the storage environment. Loss of water from produce is often associated with a loss of quality, as visual changes such as wilting or shrivelling and textural changes can take place. If using mechanical refrigeration for cooling, the larger the area of the refrigerator coils, the higher the relative humidity in the cold room will remain. It pays however, to remember that water loss may not always be undesirable, for example if produce is destined for dehydration or canning.

For fresh market produce, any method of increasing the relative humidity of the storage environment (or decreasing the vapor pressure deficit (VPD) between the commodity and its environment) will slow the rate of water loss. The best method of increasing relative humidity is to reduce temperature. Another method is to add moisture to the air around the commodity as mists, sprays, or, at last resort, by wetting the store room floor. Another way is to use vapor barriers such as waxes, polyethylene liners in boxes, coated boxes or a variety of inexpensive and recyclable packaging materials. Any added packaging materials will increase the difficulty of efficient cooling, so vented liners (about 5% of the total area of the liner) are recommended. The liner vents must line up with the package vents to facilitate cooling of the produce inside. Vented liners will decrease VPD without seriously interfering with oxygen, carbon dioxide and ethylene movement.

Room Cooling

Room cooling is a relatively low cost but slow method of cooling when electricity for mechanical refrigeration is available. The greater the refrigerator's evaporator coil area, the less moisture will be lost from the product as it cools.

It is important to leave adequate space between the stacks of boxes inside the refrigerated room in order for produce to cool more quickly. Stacks should be narrow, about one pallet width in depth. Air circulating through the room passes over surfaces and through any open space, so cooling from the outside to the center of the stacks is mostly by conduction.

Low cost cold rooms can be constructed using concrete for floors and polyurethane foam as an insulator. Building the storeroom in the shape of a cube will reduce the surface area per unit volume of storage space, thereby reducing construction and refrigeration costs. All joints should be calked and the door should have a rubber seal.

Forced-air Cooling

Forced-air cooling pulls or pushes air through the storage containers themselves, greatly speeding the cooling rate of any type of produce. Many types of forced-air coolers can be designed to move cold moist air over the commodities.

Cold wall forced-air cooler

A portable forced-air cooler can be constructed using a canvas or polyethylene sheet. The sheet is rolled over the top and down the back of the boxes to the floor, sealing off the unit and forcing air to be pulled through the vents of the cartons stacked against the cooler. This unit is designed to be used inside a refrigerated storage room.

A portable forced-air cooler

The illustration below shows the recommended pattern of vents for cartons used to hold produce that is to be forced-air cooled. Vents should make up 5% of the total surface area, and should be located 5 to 7.5 cm (2 to 3 inches) away from the corners. A few large vents (1.3 cm =0.5 inch wide or more) are better than many small vents.

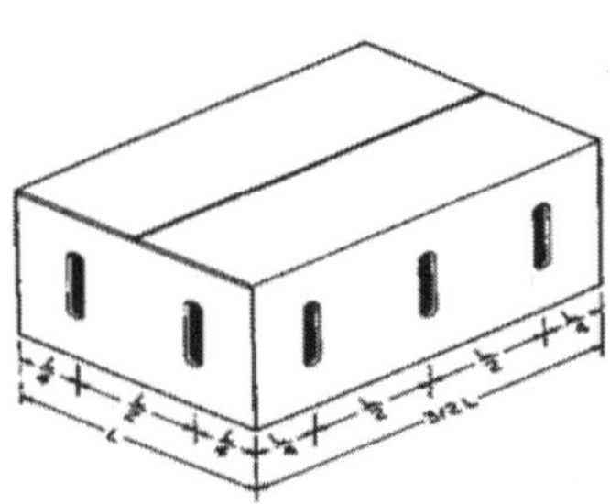

Hydro-cooling

Hydro-cooling provides fast, uniform cooling for some commodities. The commodity as well its packaging materials must be tolerant of wetting, chlorine (used to sanitize the hydro-cooling water) and water beating damage.

The simplest version of a hydro-cooler is a tank of cold water in which produce is immersed. The type shown below showers a batch of produce with icy water as the produce moves along a conveyor. A batch-type hydro-cooler can be constructed to hold entire palletloads of produce. Conveyors can be added to help control the time produce stays in contact with the cold water.

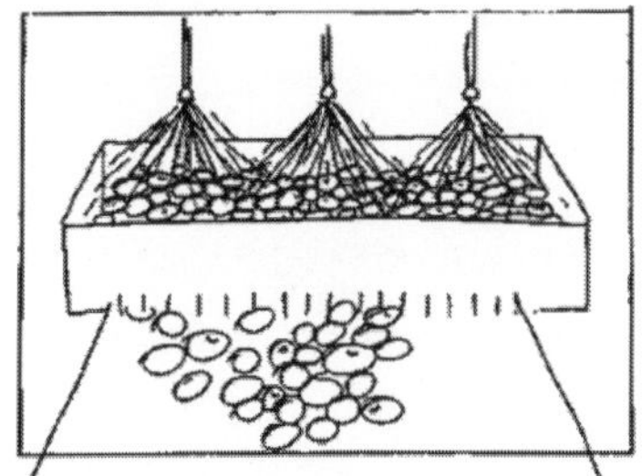

Hydro-cooling

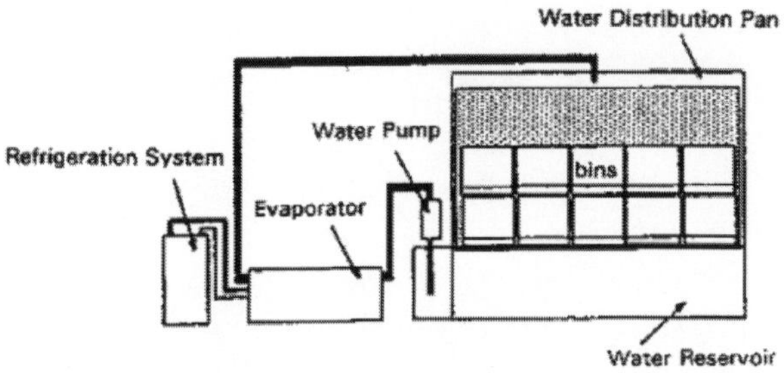

Batch-type hydro-cooler

Evaporative Cooling

These packinghouses are made from natural materials that can be moistened with water. Wetting the walls and roof first thing in the morning creates conditions for evaporative cooling of a packinghouse that is made from straw.

Straw packinghouse

The packinghouse illustrated below is made with walls of wire mesh that hold charcoal. By moistening the charcoal with water each morning, the structure will be evaporatively cooled during the day.

Evaporative coolers can be constructed to cool the air in an entire storage structure or just a few containers of produce. These coolers are best suited to lower humidity regions, since the degree of cooling is limited to 1 to 2 C (2 to 4 F) above the wet-bulb temperature. A cooling pad of wood fiber or straw is moistened and air is pulled through the pad using a small fan. In the example provided here, O. 5 gallon of water per minute is dripped onto an 8 square foot pad, providing enough moist air to cool up to 18 crates of produce in 1 to 2 hours. Water is collected in a tray at the base of the unit and recirculated.

An evaporative cooler can be combined with a forced air cooler for small lots of produce. Air is cooled by passing through the wet pad before it passes through the packages and around the produce. The air can be cooled to within a few degrees of the wet bulb temperature of ambient air.

Evaporative forced-air cooler

The evaporative cooler shown below is equipped with a vortex wind machine. Chicken wire was used to construct two thin boxes on opposite sides of the cooler that hold wet chunks of charcoal or straw. Water is dripped onto the charcoal or straw, and the wind turns the turbine, sucking moist, cool air through the load of produce inside the cooler. When using this cooler, temperatures are reduced to 3 to 5 C (6 to 10 F) below ambient air temperature, while relative humidity is about 85 %.

Drip cooler

Evaporative coolers can be constructed from simple materials, such as burlap and bamboo. The "drip cooler" shown here operates solely through the process of evaporation, without the use of a fan. Cooling will be enhanced if the unit is kept shaded and used in a well ventilated area.

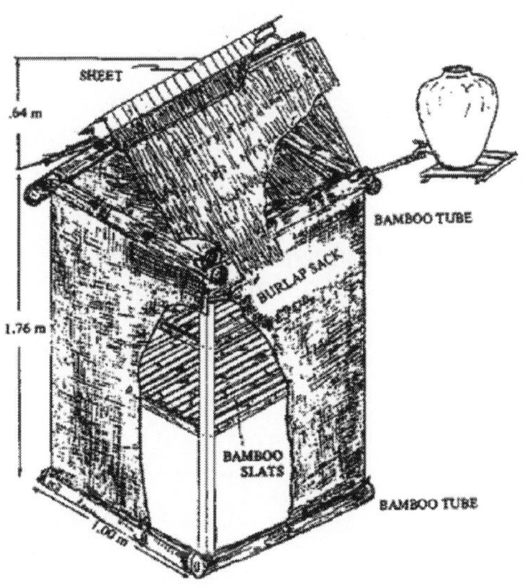

Drip cooler

The cavity between the walls is filled with sand and the bricks and sand are kept saturated with water. Fruits and vegetables are loaded inside, and the entire chamber is covered with a rush mat, which is also kept moist. Since a relatively large amount of materials are required to construct this cold storage chamber, it may be useful only when handling high value products.

During the hot summer months in India, this chamber is reported to maintain an inside temperature between 15 and 18 C (59 and 65 F) and a relative humidity of about 95%.

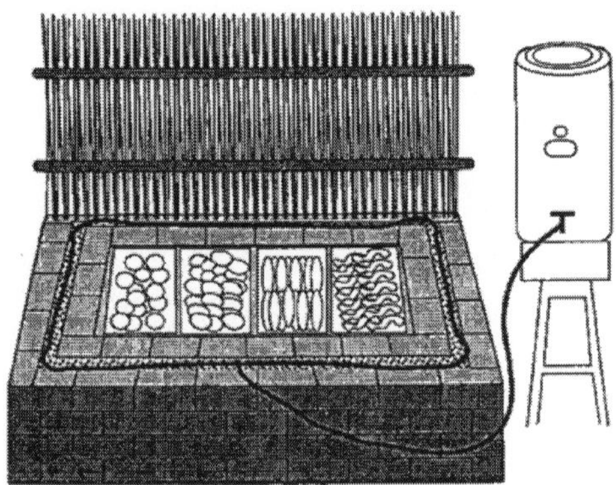

Improved Zero-Energy Cool Chamber

Night Air Ventilation

Storage structures can be cooled using night air if the difference in day and night temperature is relatively large. The storage facility should be well insulated and vents should be located at ground level. Vents can be opened at night, and fans can be used to pull cool air through the storeroom The structure will best maintain cool temperatures during the heat of the day if it is well insulated and vents are closed early in the morning.

Chilling Injury

Fruit and vegetable crops often are susceptible to chilling injury when cooled below 13 to 16 C (55 to 60 F). Chilling injury reduces the quality of the product and shortens shelf life. The table below provides some examples of the symptoms of chilling injury in a variety of crops. Symptoms often appear only after the commodity is returned to warmer temperatures, as when marketed.

Use of Ice

Ice can be used as a bunker source of refrigeration (used by passing air through a bank of ice and then through the commodity) or as top ice (laid directly in contact with the product). Ice can cool a commodity only if it melts, so good ventilation is necessary for effective cooling.

Crushed or flaked ice for package icing can be applied directly or as a slurry in water. The use of ice to cool produce provides a high relative humidity environment around the product. Package ice can be used only with water tolerant, non-chilling sensitive products (such as: carrots, sweet corn, cantaloupes, escarole, lettuce, spinach, radishes, broccoli, green onions). and with water tolerant packages (waxed fiberboard, plastic or wood).

Top ice is used for certain products during transport to help maintain a high relative humidity. Top ice can be used only with water tolerant, non-chilling sensitive products (such as: carrots, sweet corn, cantaloupes, escarole, lettuce, spinach, radishes, broccoli, green onions), and with water tolerant packages (waxed fiberboard, or wood).

Top-ice on loads should be applied in rows rather than a solid mass. It is important not to block air circulation inside the transport vehicle.

Should be Top-iced	*Can be Top-iced*
beets with tops	artichokes, globe
broccoli	beet greens
carrots with tops	beets topped
corn sweet	brussels sprouts
endive	cantaloupes
escarole	carrots, topped
green onions	celeriac
parsley	chard
radishes with tops	kohlrabi
turnips with tops	leeks
watercress	mustard greens
radish greens	parsnips
spinach	radishes
turnip greens	rutabagas

Alternative Methods of Cooling

Radiant Cooling

Radiant cooling can be used to lower the air temperature in a storage structure if a solar collector is connected to the ventilation system of the building. By using the solar collector at night, heat will be lost to the environment. Temperatures inside the structure of 4 C (about 8 F) less than night temperature can be achieved.

Use of Well Water

Well water is often much cooler than air temperature in most regions of the world. The water temperature of a deep well tends to be in the same range as the average air temperature of the same locality. Well water can be used for hydro-cooling and as a spray or mist to maintain high relative humidity in the storage environment.

High Altitude Storage

As a rule of thumb, air temperatures decrease by 10 C (18 F) for every one kilometer increase in altitude. If handlers have an option to pack and/or store commodities at higher altitude, costs could be reduced. Cooling and storage facilities operated at high altitude would require less energy than those at sea level for the same results.

Increasing relative humidity

Refrigerated air tends to be lower in relative humidity than is beneficial for storage of most horticultural crops. The simplest method of increasing relative humidity of the storage air is to wet the floor of the room or mist the storage containers with cold water and allow the water to evaporate. For a more permanent system of high relative humidity in storage environment, moisture can be added to the refrigerated air. A fan draws air past the refrigerator's evaporator coils (R) then past wet moss or straw (M). The moist air is then pulled into the store-room through a perforated wall (P).

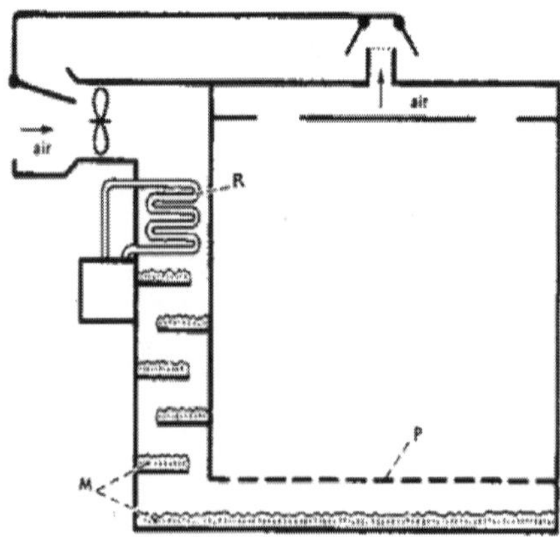

Wet moss as a moisture source inside a refrigerated storeroom:

Using a polyethylene liner in a fiberboard carton can help protect produce and reduce water loss in commodities such as cherries, nectarines, kiwifruits, bananas and herbs. Water vapor given off by the product is contained within the inner, increasing the RH around the product. The liner can also reduce abrasion damage that results from fruit rubbing against the inside of the box.

References

Brackett, R.E. 1999. Incidence, contributing factors, and control of bacterial pathogens in produce. *Postharv. Biol. Technol.* 15:305-311.

FAO. 1989. *Prevention of Post-Harvest Food Losses: Fruit. Vegetables and Root Crops.* A Training Manual. Rome: UNFAO. 157 pp.

Lamikanra, O., ed. 2002. *Fresh-cut fruits and vegetables: science, technology, and market.* Boca Raton, FL: CRC Press, 456 pp.

Proctor, F.J. 1985. Post-harvest handling of tropical fruit for export. *The Courier* 92: 83-86.

7

Storage of Horticultural Crops

If produce is to be stored, it is important to begin with a high quality product. The lot of produce must not contain damaged or diseased units, and containers must be well ventilated and strong enough to withstand stacking. In general proper storage practices include temperature control, relative humidity control, air circulation and maintenance of space between containers for adequate ventilation, and avoiding incompatible product mixes.

Commodities stored together should be capable of tolerating the same temperature, relative humidity and level of ethylene in the storage environment. High ethylene producers (such as ripe bananas, apples, cantaloupe) can stimulate physiological changes in ethylene sensitive commodities (such as lettuce, cucumbers, carrots, potatoes, sweet potatoes) leading to often undesirable color, flavor and texture changes.

Temperature management during storage can be aided by constructing square rather than rectangular buildings. Rectangular buildings have more wall area per square feet of storage space, so more heat is conducted across the walls, making them more expensive to cool. Temperature management can also be aided by shading buildings, painting storehouses white or silver to help reflect the sun's rays, or by using sprinkler systems on the roof of a building for evaporative cooling. The United Nations' Food and Agriculture Organization (FAO) recommends the use of ferrocement for the construction of storage structures in tropical regions, with thick walls to provide insulation. Facilities located at higher altitudes can be effective, since air temperature decreases as altitude increases. Increased altitude therefore can make evaporative cooling, night cooling and radiant cooling more feasible. Underground storage for citrus crops is common in Southern China, while in Northwest China, apples are stored in caves This system was widely used in the U.S. during the early pert of this century.

Certain commodities, such as onions and garlic, store better in lower relative humidity environments. Curing these crops and allowing the external layers of tissue to dry out prior to handling and storage helps to protect them from further water loss.

The air composition in the storage environment can be manipulated by increasing or decreasing the rate of ventilation (introduction of fresh air) or by using gas absorbers such as potassium permanganate or activated charcoal. Large-scale controlled or modified atmosphere storage requires complex technology and management skills, however, some simple methods are available for handling small volumes of produce.

Storage Facilities

Most fruit and vegetable production in temperate areas is seasonal. In contrast, cultivation and harvest periods are much longer in tropical and subtropical areas. Demand is year round and it is normal practice to use storage in order to ensure continuity of supply. Moreover, storage is a strategy for achieving higher returns. Produce can be held temporarily to overcome gluts thus limiting price falls or to address shortage periods when prices are high. Storage time depends on the intrinsic characteristics and perishability of the product.

Shelf life ranges from short—e.g. raspberries and other berries—to those which naturally adapt to longer storage periods—e.g. onions, potato, garlic, pumpkins, etc. Storage conditions also depend on specific product characteristics. For example, some commodities tolerate temperatures close to 0 °C such as leafy vegetables. Others, such as most tropical fruits (Table 1), cannot tolerate exposure to temperatures below 10 °C.

To optimise storage conditions, not more than one crop should be stored in the same room, unless this is for a short period of time. Sharing the same storage area can result in: differences in temperature and relative humidity conditions; chilling and ethylene sensitivity; odor contamination and other problems affecting shelf life and quality.

Generally, storage facilities are linked or integrated to packinghouses or other areas where there is a concentration of product. However, often storage can also be undertaken on-farm, either naturally or in specifically designed facilities. Even under conditions of mechanical refrigeration, location and design have an impact on system operations and efficiency. First, climate is an important factor for the location of the storage facility. For example, altitude reduces temperature by 10 °C for every 1000 meters of elevation. It also increases overall efficiency of the refrigeration equipment by facilitating heat exchange with ambient temperature, thereby reducing energy costs. Shading particularly of loading and unloading areas reduces thermal differences between field and storage temperatures.

Table 1. Recommended temperature, relative humidity and storage life for fruits and vegetables

	Crop temperature (°c)	Relative Humidity (%)	Storage Life (days)
Amaranth	0-2	95-100	10-14
Apple	-1-4	90-95	30-180
Apricot	-0.5-0	90-95	7-21
Artichoke	0	95-100	14-21
Asian pear	1	90-95	150-180
Asparagus	0-2	95-100	14-21
Atemoya	13	85-90	28-42
Avocado	3-13	85-90	14-56
Babaco	7	85-90	7-21
Banana—Plantain	13-15	90-95	7-28
Barbados cherry	0	85-90	49-56
Basil	7-10	85-95	7
Bean (dry)	4-10	40-50	180-300
Beet (bunched)	0	98-100	10-14
Beet (topped)	0	98-100	120-180
Belgian endive	0-3	95-98	14-28
Blackberry	-0.5-0	90-95	2-3
Black sapote	13-15	85-90	14-21
Blueberries	-0.5-0	90-95	14
Bok Choy	0	95-100	21
Breadfruit	13-15	85-90	14-42
Broadbeans	0-2	90-98	7-14
Broccoli	0	95-100	14-21

Brussels sprouts	0	95-100	21-35
Cabbage	0	98-100	150-180
Cactus leaves	2-4	90-95	14-21
Caimito	3	90	21
Calamondin	9-10	90	14
Cantaloupe (half slip)	2-5	95	15
Cantalupo (full slip)	0-2	95	5-14
Carambola	9-10	85-90	21-28
Carrot (bunched)	0	95-100	14
Carrot (topped)	0	98-100	210-270
Cassava	0-5	85-96	30-60
Cashew apple	0-2	85-90	35
Cauliflower	0	95-98	21-28
Celery	0	98-100	30-90
Celeriac	0	97-99	180-240
Chayote	7	85-90	28-42
Cherimoya	13	90-95	14-28
Cherries	-1-0.5	90-95	14-21
Chicory	0	95-100	14-21
Chinese cabbage	0	95-100	60-90
Chives	0	95-100	14-21
Coconut	0-1.5	80-85	30-60
Cranberries	2-4	90-95	60-120
Cucumber	10-13	95	10-14
Currants	-0,5-0	90-95	7-28
Custard apple	5-7	85-90	28-42
Daikon	0-1	95-100	120
Dates	-18-0	75	180-360
Durian	4-6	85-90	42-56
Eggplant	8-12	90-95	7
Escarole	0	95-100	14-21
Fennel	0-2	90-95	14-21
Feijoa	5-10	90	14-21
Fig	-0.5-0	85-90	7-10
Garlic	0	65-70	180-210
Ginger	13	65	180
Grape	-0.5-0	90-95	14-56
Grapefruit	10-15	85-90	42-56
Green onions	0	95-100	21-28

Guanabana	13	85-90	7-14
Guava	5-10	90	14-21
Horseradish	-1-0	98-100	300-360
Husk tomato	13-15	85-90	21
Jaboticaba	13-15	90-95	2-3
Jackfruit	13	85-90	14-42
Jerusalem artichoke	-0.5-0	90-95	120-150
Jicama	13-18	65-70	30-60
Kale	0	95-100	10-14
Kiwano	10-15	90	180
Kiwifruit	-0.5-0	90-95	90-150
Kohlrabi	0	98-100	60-90
Kumquat	4	90-95	14-28
Leek	0	95-100	60-90
Lemon	10-13	85-90	30-180
Lettuce	0-2	98-100	14-21
Lima bean	3-5	95	5-7
Lime	9-10	85-90	42-56
Longan	1-2	90-95	21-35
Loquat	0	90	21
Lychee	1-2	90-95	21-35
Malanga	7	70-80	90
Mamey	13-18	85-95	14-42
Mandarin	4-7	90-95	14-28
Mango	13	90-95	14-21
Mangosteen	13	85-90	14-28
Melon (Others)	7-10	90-95	12-21
Mushrooms	0-1.5	95	5-7
Nectarine	-0.5-0	90-95	14-28
Okra	7-10	90-95	7-10
Onions (dry)	0	65-70	30-240
Olives, fresh	5-10	85-90	28-42
Orange	0-9	85-90	56-84
Papaya	7-13	85-90	7-21
Parsley	0	95-100	30-60
Parsnip	0	95-100	120-180
Passionfruit	7-10	85-90	21-35
Peach	-0.5-0	90-95	14-28
Pear	-1.5-0.5	90-95	60-210

Peas	0	95-98	7-14
Cucumber	5-10	95	28
Pepper (bell)	7-13	90-95	14-21
Persimmon	-1	90	90-120
Pineapple	7-13	85-90	14-28
Pitaya	6-8	85-95	14-21
Plum	-0.5-0	90-95	14-35
Pomegranate	5	90-95	60-90
Potato (early)	7-16	90-95	10-14
Potato (late)	4.5-13	90-95	150-300
Prickly pear	2-4	90-95	21
Pumpkins	10-15	50-70	60-160
Quince	-0.5-0	90	60-90
Radichio	0-1	95-100	14-21
Radish	0	95-100	21-28
Rambutan	10-12	90-95	7-21
Raspberries	-0.5-0	90-95	2-3
Rhubarb	0	95-100	14-28
Rutabaga	0	98-100	120-180
Salsify	0	95-100	60-120
Sapodillla	15-20	85-90	14-21
Scorzonera	0	95-98	180
Snapbeans	4-7	95	7-10
Snowpeas	0-1	90-95	7-14
Spinach	0	95-100	10-14
Sprouts	0	95-100	7
Strawberry	0-0.5	90-95	5-7
Sweet corn	0-1.5	95-98	5-8
Sweet potato	13-15	85-90	120-210
Swiss chard	0	95-100	10-14
Summer squash	5-10	95	7-14
Tomato (red)	8-10	90-95	8-10
Tree tomato	3-4	85-90	21-28
Turnip	0	90-95	120
Watercress	0	95-100	14-21
Watermelon	10-15	90	14-21
White sapote	19-21	85-90	14-21
Yam	16	70-80	60-210
Yellow sapote	13-15	85-90	21

Building design is an important factor to be taken into consideration. For example, a square shaped floor perimeter is thermally more efficient than a rectangular one. The roof is the most important part of the structure. This is because it has to protect produce from rain and radiant heat. Its slope should allow easy fall off of rainwater; its dimensions should exceed the perimeter of the building to protect walls from the sun and provide a dry area around the building in rainy weather. Floors should be of concrete, isolated from soil humidity, and elevated to avoid penetration of water.

Doors need to be wide enough for mechanised handling. Storage facilities should be thoroughly cleaned before filling. This includes brushing and washing of walls and floors to eliminate dirt and organic debris that could harbour insects and diseases. Before product is placed in the storage room, inspection and presorting should be undertaken. This is in order to remove all potential sources of contamination for the remaining load. Product should be stacked in such a way that there is free circulation of air. During storage, it should also be possible to carry out quality control inspections. If the storage facility becomes full during a long harvest period, it needs to be organised around the principles of the system "first in first out".

Storage Practices

Inspecting stored produce and cleaning storage structures on a regular basis will help reduce losses by minimizing the buildup of pests and discouraging the spread of diseases.Storage facilities should be protected from rodents by keeping the immediate area clean, free from trash and weeds. Rat guards can be made from simple materials such as old tin cans or pieces of sheet metal fashioned to fit the extended legs of storage structures. If desired, more elaborate technologies can be used. Concrete floors will help prevent rodent entry, as will screens on windows, vents and drains.

When inspecting stored produce, any spoiled or infected produce should be removed and destroyed. In some cases, produce may still be fit for consumption if used immediately perhaps as animal feed. Reusable containers and sacks should be disinfected before reuse.

Placing materials on the floor beneath sacks or cartons of produce prevents dampness from reaching produce suited to dry conditions in storage. This helps to reduce the chance of fungal infection, while also improving ventilation and/or sanitation in the storeroom. Some examples of useful materials follow:

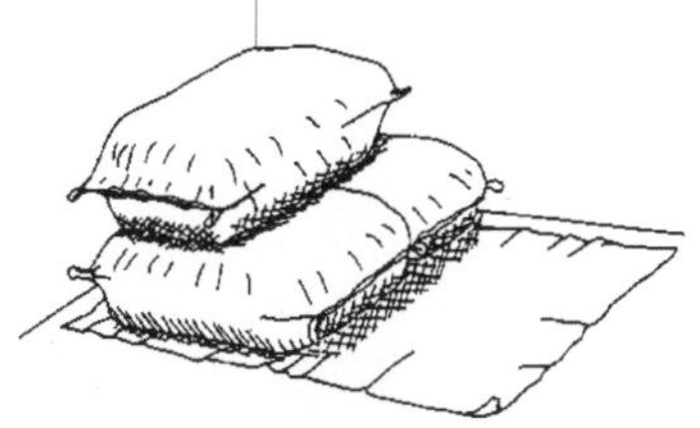

Waterproof sheets

Poles

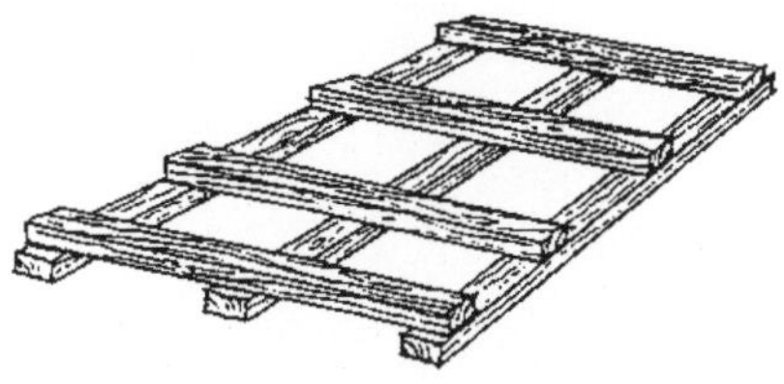

Wooden pallets

STORAGE STRUCTURES

A yam barn is a traditional structure used in West Africa to store yams after curing. Fast-growing, live trees are used to create a rectangular structure, and form the framework of the barn as well as provide shade.

Outside view of barn with 'live' shade

Trunk of fast growing tree-planted in situ.

Storage facilities require adequate ventilation to help extend shelf life and maintain produce quality. Following are the types of fans commonly used.

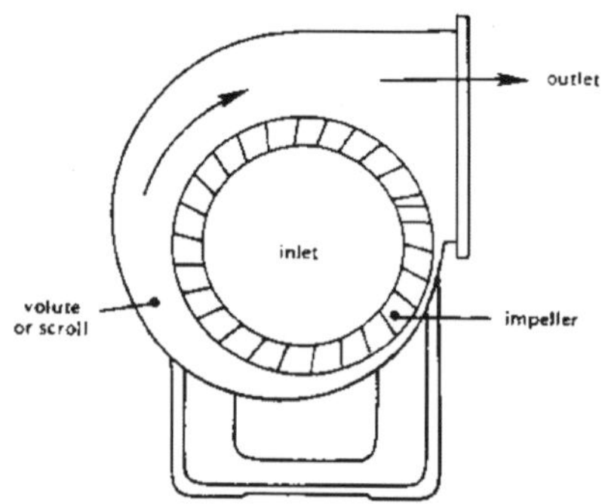

Centrifugal

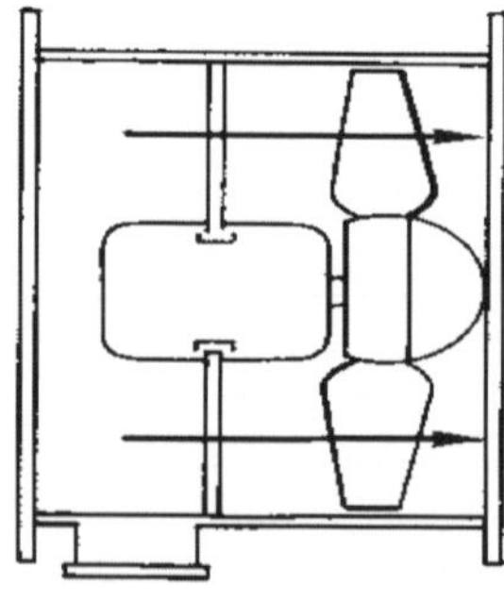

Axial flow

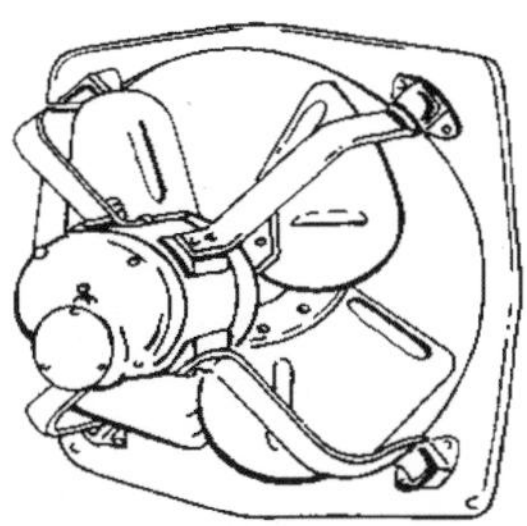

Propeller/expeller

Any type of building or facility used for storage of horticultural crops should be insulated for maximum effectiveness. A well insulated refrigerated building will require less electricity to keep produce cool.

Insulation R-values are listed below for some common building materials. R refers to resistance, and the higher the R-value, the higher the material's resistance to heat conduction and the better the insulating property of the material.

Insulation R-values

Material	*1 inch thick*
Batt and Blanket Insulation	
Glass wool, mineral wool, or fiberglass	3.50
Fill-Type Insulation	
Cellulose	3.50
Glass or mineral wool	2.50-3.00
Vermiculite	2.20
Wood shavings or sawdust	2.22
Rigid Insulation	
Plain expanded extruded polystyrene	5.00
Expanded rubber	4.55
Expanded polystyrene molded beads	3.57
Aged expanded polyurethane	6.25
Glass fiber	4.00
Polyisocyranuate	8.00
Wood or cane fiber board	2.50
Foamed-in-Place Insulation	
Sprayed expanded urethane	6.25

Building Materials	*Full thickness of material*
Solid concrete	0.08
8-inch concrete block, open core	1.11
8-inch lightweight concrete block open core	2.00
8-inch concrete block with vermiculite in core	5.03
Lumber, fir or pine	1.25
Metal siding	<0.01
3/8-inch plywood	1.25 - 0.47
1/2-inch plywood	1.25 - 0.62
Masonite particleboard	1.06
25/32-inch insulated sheathing	2.06
1/2-inch Sheetrock	0.45
1/2-inch wood lapsiding	0.81

An evaporative cooler located in the peak of a storage structure can cool an entire room of stored produce such as sweetpotatoes or other chilling sensitive crops. The vents for outside air should be located at the base of the building so that cool air is circulated throughout the room before it can exit.

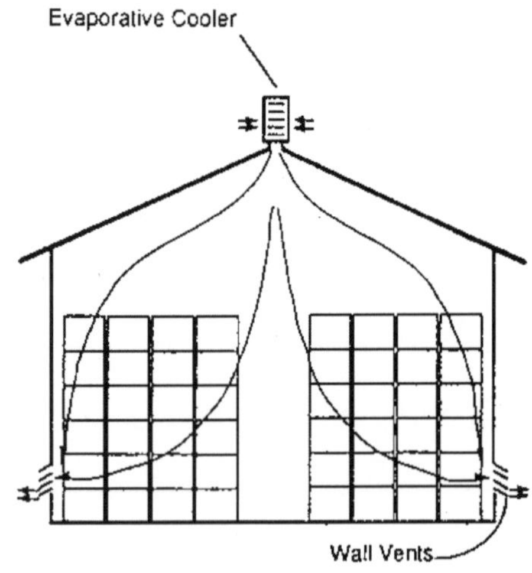

Commercially constructed cold rooms can be quite expensive, but fortunately the small-scale operator has many choices. Cold rooms can be self-constructed, purchased as prefabricated units, or made from refrigerated transportation equipment such as railway cars, vans or marine containers.

For storage facilities that are refrigerated, using outside air for ventilation is wasteful of energy. For these systems, a simple recirculation system can be designed by adding a fan below floor level and providing a free space at one end of the storeroom for cool air to return to the inlet vents.

The proper arrangement of floor vents for air circulation will improve ventilation in the storage house. Lateral ducts should be 2 meters apart when measured from center to center. Air How velocities from the main duct should be 10 to 13 meters/second.

Storage structures can be cooled by ventilating at night when outside air is cool. For best results, air vents should be located at the base of the storage structure. An exhaust fan located at the top of the structure pulls the cool air through the storeroom. Vents should be closed at sunrise, and remain closed during the heat of the day.

Overhanging roof extensions on storage structures are very helpful in shading the walls and ventilation openings from the sun's rays, and in providing protection from rain. An overhang of at least 1 meter (3 feet) is recommended. Where electricity is not available, wind-powered turbines can help keep storerooms cool by pulling air up through the building. Vents at the floor level are especially useful for cooling via night air ventilation.

The turbine illustrated below can be constructed of sheet metal that is twisted to catch the wind, and attached to a central pole that acts as the axis of rotation. Warm air in the storage room rises, causing the turbine to rotate, expelling the air and initiating an upward flow of warm air. The turbine should be placed on the peak of the roof of a storage structure.

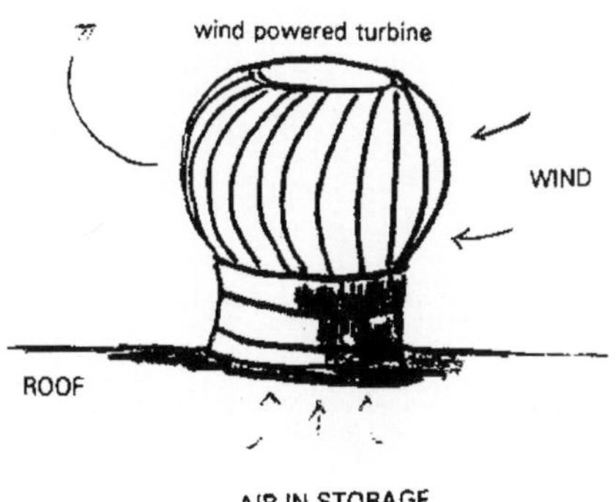

One of the simplest methods for storing small quantities of produce is to use any available container, and create a cool environment for storage by burying the container using insulating materials and soil.

Dried and Bulb Crops

Onions, garlic and dried produce are best suited to low humidity in storage. Onions and garlic will sprout if stored at intermediate temperatures. Pungent types of onions will store longer than mild onions, which are rarely stored for more than one mont. The following table lists the storage conditions recommended for these crops.

	Temperature		*RH*	*Potential storage duration*
	C	*F*	*%*	
Onions	0-5	32-41	65-70	6-8 months
	28-30	82-86	65-70	1 month
Garlic	0	32	70	6-7 months
	28-30	82-86	70	1 month
Dried fruits and vegetables	<10	<50	55-60	6-12 months

For bulk storage of onions or garlic, ventilation systems should be designed to provide air into the store from the bottom of the room at a rate of 2 cubic feet per minute per cubic feet of produce. If produce is in cartons or bins, stacks must allow free movement of air. Rows of containers should be stacked parallel to the direction of the flow of air and be spaced six to seven inches apart. An adequate air supply must be provided at the bottom of each row and containers must be properly vented.

Root and Tuber Crops

The recommended storage conditions for root and tuber crops are listed in the following table. Potatoes for processing are best kept at intermediate temperatures to limit the production of sugars which darken when heated during processing. Potatoes meant for consumption must also be stored in the dark, since the tubers will produce chlorophyll (turning green) and develop the toxic alkaloid solanine if kept in the light. Potatoes stored for use as "seed" are best stored in diffuse light. The chlorophyll and solanine that accumulate will aid to protect the seed potatoes from insect pests and decay organisms.

	Temperature		*RH(%)*	*Potential storage duration*
	C	*F*		
Potatoes				
Fresh market	4-7	39-45	95-98	10 months
Processing	8-12	47-54	95-98	10 months
Seed potatoes	0-2	32-36	95-98	10 months
Cassava	5-8	41-47	80-90	2-4 weeks
	0-5	32-41	85-95	6 months
Sweetpotato	12-14	54-58	85-90	6 months
Yam	13-15	55-59	near 100	6 months
	27-30	80-86	60-70	3-5 weeks
Ginger	12-14	54-58	65-75	6 months
Jicama	12-15	54-59	65-75	3 months
Taro	13-15	55-59	85-90	4 months

Tropical root and tuber crops must be stored at temperatures that will protect the crops from chilling, since chilling injury can cause internal browning, surface pitting and increased susceptibility to decay.

Potatoes

When storing potatoes, a field storage clamp is a low cost technology that can be designed using locally available materials for ventilation and insulation. The example illustrated here employs a wooden ventilator box and straw for insulation. The entire pile of potatoes and straw is covered with a layer of soil, which should not be highly compacted. In very cold regions, a second layer of straw and soil can be added. In hot regions, less soil is needed, but more ventilation can be added by constructing chimney type air outlets at the top of the clamp.

Simple storage houses for potatoes can be constructed for small quantities of produce. The examples provided here can store 1 to 2 metric tons, and are used on farms and in mountain villages. The first is made from unfinished wooden planks painted white to reduce heat accumulation from the sun and covered with a large thatched roof for protection from sun and rain. It has a large door on one side for loading and unloading.

The second storage house is constructed from lath and plaster and mud bricks in a cylindrical form. It has two doors, one on top for loading, the other at the bottom for easy removal of potatoes for sale or consumption. White-wash helps reduce heat accumulation and a thatch roof protects the potatoes from rain and sun.

For large quantities of potatoes, a self-supporting A-frame storehouse can be constructed. A pit is dug about 10 feet deep and wooden air ducts are placed along the earthen floor. The roof of the building is constructed of wood, then covered with straw and soil.

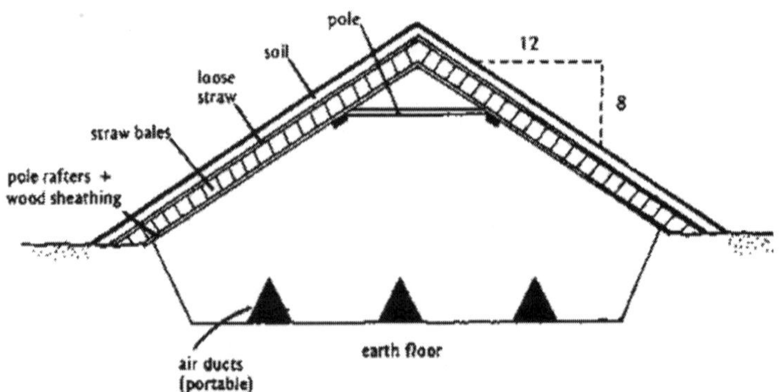

Ducts for ventilation of bulk storage rooms can be laid out vertically as well as horizontally. The storeroom for potatoes shown below provides for plenty of ventilation using simple materials. The room can be of any size or shape since air ducts can be positioned to extend evenly throughout.

When loading potatoes into bulk storage, even distribution of the produce is important for proper ventilation. Uneven loads will inhibit air movement and result in storage losses due to inadequate ventilation.

Short-term Storage

Refrigeration in cold stores is not always used to maximise the postharvest life. On the contrary, it is probably used more often during the short time required for the sequence of activites in the cold chain ending at the consumption point. Refrigerated transport is probably the best example of this. However, there are many other opportunities for the use of temporary cold storage e.g. during preparation and sale of product for the market. For example, holding product until processing, packaging, or transport is carried out. Other examples include the use of refrigerated facilities at wholesale or retail.

Cold storage is also used in the home to prolong the shelf-life of products. It is hard to define what constitutes "short term and long-term storage". This is because 7 days is a long time for raspberries while for potato, onion, garlic and other products that require longer periods of storage this is considered to be relatively short.

It is preferable not to store different crops together. However, this is common practice and is unavoidable in many cases, particularly at distribution or retail. This does not pose a problem provided products are not exposed to suboptimal conditions for too long and build up of ethylene is avoided. A strategy widely practiced is to set cold chambers at an average of around 5°C and 90-95% level of relative humidity. If possible, mixed loads should have different regimes depending on the specific combination of fruits and vegetables in store. This is assuming that ambient ethylene concentration does not exceed 1 ppm.

Transport is an example of temporary refrigerated storage. Mixed loads cause incompatibility problems highlighted previously. Because packaging dimensions are different, they are usually not fully stackable and the ventilation openings of packages of different dimensions do not match to each other. This prevents ventilation and creates microambient conditions which are undesirable.

Facilities for long-term storage of potato, onion, sweet potato etc, often include using a combination of forced air systems as well as heating and/or refrigeration equipment. Because these are crops that initially require a curing period, hot and humid air is introduced at the beginning. Later, the temperature is reduced with either through forced air cooling or natural ventilation. Adequate temperatures are obtained by mixing external and internal atmospheres and if required, the air is heated or refrigerated. In this way, the same building is used for both curing and storage—an important consideration in mechanised harvesting systems.

Controlled Atmospheres

With atmosphere modification, the low metabolic rate achieved with refrigeration is extended even further. As a result, the storage period is prolonged without further losses in quality. Composition of normal atmosphere at sea level is around 78,1% nitrogen, 21% oxygen y 0,03% carbon dioxide. A "controlled" or "modified" atmosphere is obtained when its composition varies from the norm.

In controlled atmosphere, gas composition is exactly maintained. It is often used for extremely long periods of storage in purpose built facilities. Modified atmospheres, on the other hand, are obtained when produce is packed in semi permeable films and are used for short periods. The atmospheric composition inside the package changes until it is in equilibrium with the ambient one. Equilibrium atmosphere depends on product, film characteristics, and storage temperature.

The modification of storage atmosphere delays the biochemical and physiological changes associated with senescence. This mainly involves the respiratory rate, ethylene production, softening and compositional changes. Other effects include the reduction in sensitivity to ethylene, and in some cases chilling and the severity of pathogen attack. The atmospheric composition can also be used to control insects. The risk of using abnormal atmospheres is that they may cause fermentation, tissue asphyxia, and the development of off-odors or off-flavours. From the construction point of view, controlled atmosphere facilities are similar to refrigeration facilities.

However, they should be airtight to allow creation of an atmosphere different from normal. The Oxygen consumption and its replacement by carbon dioxide by respiration, create the atmosphere. When the appropriate combination has been reached, a limited intake of oxygen is required to satisfy the reduced rate of respiration. Accumulation of carbon dioxide is removed by means of different methods. Because internal atmosphere behaves differently, a pressure compensating system is required to attain equilibrium with the external or ambient atmosphere.

As controlled atmosphere rooms are kept locked until the end of the storage period, inspection windows are required to control refrigeration equipment. Product should also be placed at the top of one of the walls. Atmospheric composition is crop specific. However, as a general rule the most common combinations are 2-5% oxygen and 3-10% carbon dioxide. Many crops benefit from atmosphere modification.

However, usage is limited. It is difficult to define products ideal for storing under controlled atmosphere. However, one of the most important factors is that investment and operating costs should be recovered. Other factors include: First, products should be seasonal and have a stable demand during a long marketing period. Second, product should have some unique qualities and not be easily substituted by similar products. In other words, it is beneficial to use controlled atmosphere technology when there are no competitor products on the market. This may go some way towards explaining why its usage is limited to specific crops.

Home Storage System

Outdoor Storage

Produce that requires cool-to-cold moist surroundings can be stored outdoors. All outdoor storage has the disadvantage of being unaccessible sometimes and subject to damage by rodents and other vermin. A welldrained location is essential to prevent excessive accumulation of water. Usually the produce must be insulated for protection from frost and fluctuating temperatures. Insulating materials commonly used are straw, hay, dry leaves, corn stalks, or wood shavings, and some soil. Be sure that the insulating materials used are not contaminated with pesticides.

In-garden Storage

It is possible to leave some root crops, such as carrots, turnips, and parsnips in the garden where they grew, for part or all of the winter. After the ground begins to freeze in the late fall, cover the root crops with a foot or more of mulch-straw, hay, or dry leaves. Do not place mulch on warm soil because doing so will cause vegetables to decay rapidly.

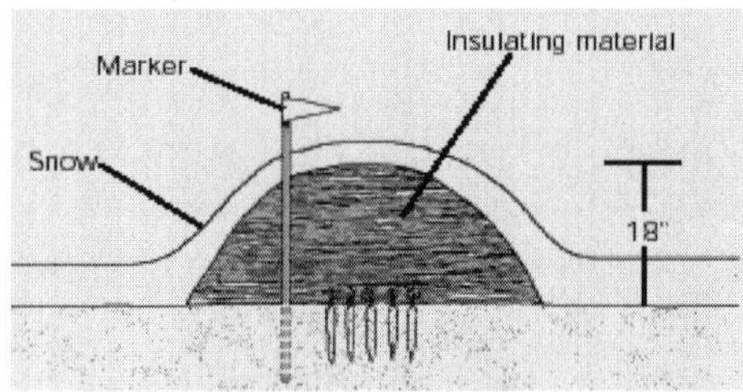

In-garden storage

Produce can be difficult to dig out of the frozen ground, but it will not be adversely affected until the temperature around the roots drops to 25° or less. Carrots are damaged at about 25°, but parsnips can stand somewhat lower temperatures. If rodents are a problem, it may be wise to store produce in a buried container or an indoor storage area. One gopher can consume a whole row of carrots left in the ground. Parsnips, horseradish, and turnips actually improve in flavour by light freezing. At temperatures between 28° and 34°, the starch changes to sugar. Other crops, such as beets, cabbage, Chinese cabbage, cauliflower, celery and onions can withstand the early light frosts and can be stored for several weeks under a heavy mulch.

Mounds, pits

Mounds or pits are a very economical way to store cabbage and root crops, such as carrots, beets, celeriac, kohlrabi, rutabagas, turnips, and winter radishes. Select a well-drained location, and cover the ground with an insulating mulch.

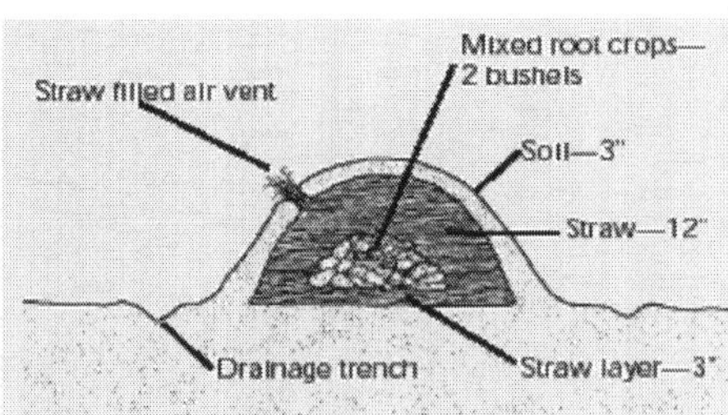

Mound storage

Making a shallow excavation (from 6 to 10 inches deep) and placing the produce partly below the surface will ensure better frost protection, but it will also increase the danger of excess water. Place mulch over vegetables. A ditch around the storage perimeter will help remove surface water.

Vegetables keep very well in pits and mounds, but once these storage areas are opened all the produce should be removed. After it removed, the produce will keep for 1 or 2 weeks at most. It does not keep as long after removal from storage as will freshly harvested produce. Root crops can be mixed, but should be separated with mulch to prevent cross-transfer of odors.

Buried containers

A 20-gallon trash can, buried in the ground, makes a convenient and economical storage for many kinds of vegetables. Metal cans are more rodent proof than plastic.

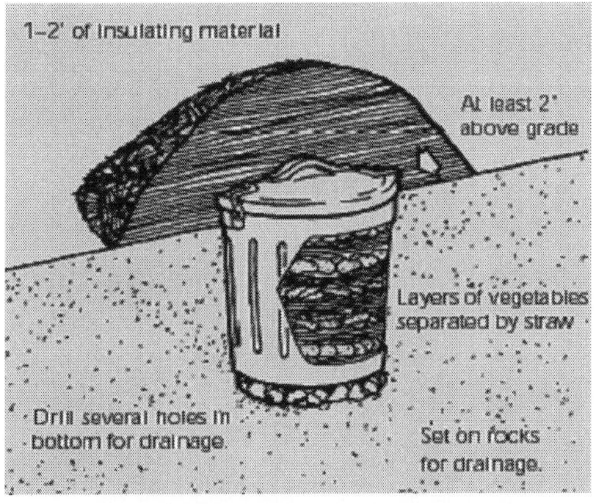

Buried trash can storage.

Several holes should be made in the bottom to facilitate drainage. The container must be free of substances that might give off-flavour to the produce. Never use drums or containers that might have held pesticides or other chemicals.

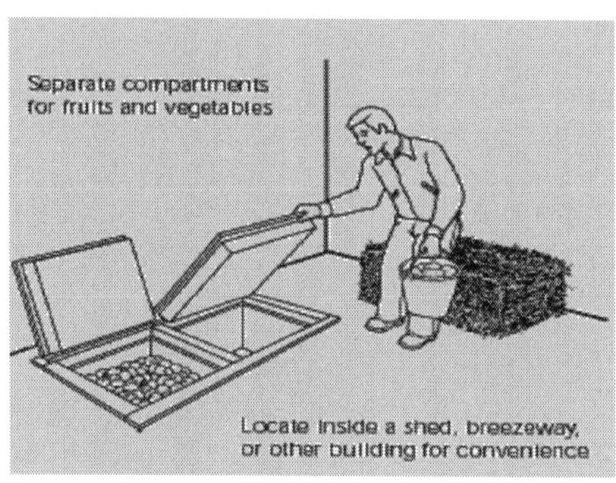

Buried insulated box storage

An insulated box or other container can be buried in a well-drained area. Buried containers are more easily opened and closed than mounds and trenches. This type of storage could be located in a breezeway, shed, or garage for easier access and greater frost protection. However, if you plan to store food in or near a garage, you must more carefully wrap and protect the produce from car fumes, which produce easily picks up.

Indoor Storage Area

There are many areas in dwellings that naturally provide, or can be adapted to provide a variety of temperature and moisture conditions for storage. These include the attic, unheated rooms, the basement, or cellar.

Assess your own storage possibilities. Use a thermometer to monitor temperatures. Any spot that is sufficiently and evenly cool can be adapted for some type of food storage. The relative humidity of these locations will also affect what type of produce can be stored.

Even basement window wells can be converted to small storage areas. They can provide cool, moderately moist conditions and can be used to store some types of apples or a variety of root crops. But be sure to separate the produce properly. Insulate produce with bales of hay or straw. A pantry or unheated room is useful for short-term storage of potatoes and onions, and long-term storage of spices, vegetable oils, nuts, and commercially canned goods. Low storage temperatures extend the shelf life of dried foods, such as dried beans, herbs, dried fruits and vegetables, and the life of other products, such as coffee, flour, rice, pasta, and cereals. Use containers with tight-fitting lids to keep humidity low and rodents and insects out. Guard against freezing when storing in unheated areas. A warm storage area, such as an attic, can be a good environment in the fall for drying herbs, beans, walnuts, or hickory nuts.

Basement storage

A well-ventilated basement with central heating is generally dry and has a temperature range of 50° to 60°. It may be used for ripening tomatoes and for short-term storage of pumpkins, winter squash, potatoes, sweet potatoes, and onions. If you will be storing only a small amount of produce, an old refrigerator placed in one corner of the basement may be an excellent investment. It can be used for storing head lettuce, cauliflower or cabbage, or apples.

For long-term storage over winter you will need to partition off a room and insulate it to control temperature and moisture. Build a storage room on the north or east side of the basement, if practicable, and make sure there are no heating ducts or pipes running through it. You will need at least one window for cooling and ventilating the room. Two or more windows are desirable, particularly if the room is divided for separate storage of fruits and vegetables.

A room 4 x 4 x 6 feet is large enough for most households. An area 6 x 7 x 7 feet will hold approximately 30 bushels of produce—5 bushels per month for 6 months. An area 8 x 10 x 7 feet will hold approximately 60 bushels of produce—10 bushels per month for 6 months. To deter rodents, carefully install sheathings or screens on the partition walls so there are no openings at the floor or ceiling. Young mice can crawl through a quarter-inch hole.

Storage Area Management

Temperature Management

Proper temperatures are critical to long-term storage of vegetables and fruits. Improper temperatures, whether too cool or too warm, can cause rapid deterioration of produce quality. Two reliable thermometers (preferably that record minimum and maximum temperatures) should be used to observe and regulate storage temperatures carefully. Place one thermometer in the coldest location of the basement or cellar, and place the other thermometer outdoors. Regulate storage temperature by opening and closing doors, windows, or other openings used as ventilators.

Outdoor temperatures well below 32° are necessary to cool storage air to 32° and to maintain that temperature. Once cooled to 32°, the temperature will rise again if ventilators are closed, even when outdoor temperature is about 25°. Close ventilators tightly whenever the outdoor temperature is higher than the storage temperature. Watch indoor and outdoor temperatures closely. In most regions, daily adjustment of ventilators is usually necessary to maintain desired storage temperatures.

The danger of overventilating during subfreezing weather is that stored products will freeze if you are not careful. For example, in a study made of an insulated storage cellar partly above ground, full ventilation both day and night was necessary to maintain a temperature of 32° if outdoor temperatures ranged between 18° and 30° during the day and dipped to 10° at night. If minimum temperature at night, with normal air movement, was 8° for 5 or 6 hours, the storage temperature dropped to 30°. During a night of high wind, however, a minimum of 12° cooled the cellar to 30°. Keep a pail of water in the storage area in very cold weather; water will freeze before the crops. However, very low temperatures can cause chilling injury to some produce, like potatoes, and leave them unusable or increase decay.

Managing Humidity Level

The proper humidity level is also very important to long-term storage and maintenance of good quality produce. This is generally measured as relative humidity, which is the percentage of moisture in the atmosphere at a given time as related to the maximum amount (100%) that could be retained at that temperature. A humidity level of 95% is almost rainfall and rather difficult to achieve indoors. A relative humidity of 90%—5% is very moist and good for storage of potatoes and other root crops.

A relative humidity of 60%—5% is dry and good for storage of pumpkins and other squash. Two ways of maintaining proper humidity are first, to use water to raise the humidity of the storage air, and secondly, to use ventilated polyethylene bags and box liners. Moisture can be added to storage air by sprinkling the floor frequently, by placing large pans of water under fresh-air intake vents, by covering the floor with wet materials such as straw or odorless sawdust, or by a combination of these methods. However, these strategies will not prevent shrivelling of root crops.

The easiest and most effective way to control moisture loss in root crops and in certain other crops is to put them in polyethylene bags or box liners. Cut many ¼- to 3/8-inch holes in the sides of the bags or liners to permit ventilation. Tie the bags and fold over the tops of box liners, but do not seal them. Excessive humidity is conducive to mold and decay organisms, if water droplets are allowed to form on the surface of the product. Plastic bags and liners used for moisture retention should always be perforated at regular intervals to permit air circulation and prevent condensation.

Sanitation

At least once a year, remove all containers from the storage area and clean and air them in the sun. The room itself should also be thoroughly aired, cleaned and washed down with a disinfectant, such as diluted chlorine bleach (1 cup bleach mixed with 9 cups water), to kill off any molds or bacteria that could lie dormant and ruin future crops.

References

Gorny, J.R., ed. 2001. *Food safety guidelines for the fresh-cut produce industry*. Fourth edition. Alexandria, VA: International Fresh-cut Produce Association. 218 pp.

Kader, A.A., ed. 2002. *Post-harvest technology of horticultural crops*. Oakland: University of California, Division of Agriculture and Natural Resources Publication 3311, 535 pp.

McGregor, B.M. 1989. *Tropical products transport handbook.* USDA Agr. Handb. No. 668, 148 pp.

Paull, R.E. and J.W. Armstrong, eds. 1994. *Insect pests and fresh horticultural products: treatments and responses.* Wallingford, UK: CAB International, 360 pp.

8

Transportation of Horticultural Crops

Transportation is often the most costly factor in the marketing channel, and for airfreighted export crops the cost of transportation may exceed the cost of production. The method of transportation for fresh fruit and vegetables is determined by distance, perishability and the value of the product. Whatever the method used, the principles of transport are the same:

- Loading and unloading should be as careful as possible.
- Transmit times should be as short as possible.
- The product should be well protected in relation to its susceptibility to physical injury.
- Jolting and movement should be reduced as much as possible.
- Overheating should be avoided.
- Water loss by the produce should be restricted.
- The required conditions of preservation should be obtained and maintained constantly, regarding in particular temperature, relative humidity and air circulation.

Handling and Moving Short Distances

Marketing and physical distribution of fresh produce inherently means moving the produce. The commodities are handled, either manually or mechanically, many times from harvest and through the distribution process before the consumer buys and prepares them to eat.

Handling operations are seldom given much thought by the individuals directly involved in moving the process, particularly when the produce is only moved short distances.

Some useful equipment for moving produce short distances which would not only make the job much easier for the workers are shown below. It would also speed up operations and reduce physical injury to the produce at the same time.

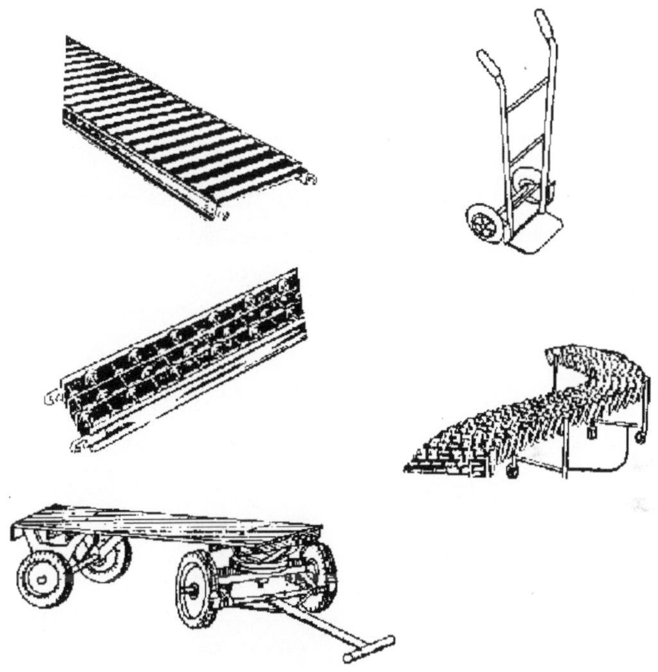

Equipment for moving fresh produce short distances

Roller-conveyors, be they motorised or gravity fed are of great assistance in the packing house for moving boxed produce but are equally useful for loading and unloading of pickups, trucks and stores. Hand-trucks can carry up to six or more crates of produce, are very manoeuvrable light and durable, and do not cost a lot of money. Hand-carts with the front wheels steered by the tow-bar can carry a lot more produce but need to be used on more level ground. Pallet-trucks are in everyday use in the Eastern Caribbean but are not yet used with any regularity for fresh produce handling and movement, probably because packaging has yet to be sufficiently standardised to benefit from pallet operations.

Handling and Stowage during Transportation

Dropping of packages during loading and unloading is a frequent cause of damage to the produce and to the package, but can be minimised by:

— Using pack weights and designs which are compatible with the handling method.

— Correct supervision and management of loading and unloading to prevent careless handling and to ensure workers are strong enough and tall enough for the Job.

— Using ramped loading bays gives a tremendous advantage when loading trucks with produce.

— Providing shelter from sun and rain at loading and unloading areas.

— Using trolleys, conveyors and fork-lift trucks to reduce the amount of manual handling.

The method of stowage of the produce in the transport vehicle will depend on the pack, the commodity and the type and size of the vehicle but should always be carefully planned and managed to minimise both physical and environmental damage. The following are some useful guidelines:

— Loading tightly to reduce movement and make best use of space.

— Distribute weight evenly.

— In mixed commodity or dispatch loads, stow goodsin reverse order to their unloading sequence.

— Provides gaps for ventilation.

— Only stack to a load height which the lowest containers can withstand without crushing or damage to the pack.

— Do not exceed the capacity of the vehicle.

— Ensure the vehicle is properly maintained, breakdowns are time delaying and may lead to excessive or total spoilage of the produce.

— Be selective as regards the vehicle operator, e.g. band and inexperienced drivers will mean more damage to the produce and to the vehicle.

Road Transportation

Inland transportation of horticultural crops is usually by rail or way truck. Overseas transportation is by sea or by air. A limited amount of high-valued produce is sometimes transported overland by air. The basic requirements for conditions during transportation are similar to those needed for storage, including proper control of temperature and humidity and adequate ventilation. In addition, the produce should be immobilised by proper packaging and stacking, to avoid excessive movement or vibration. Vibration and impact during transportation may cause severe bruising or other types of mechanical injury.

CA (Controlled Atmosphere) containers are not used much for transporting fruit and vegetables. Only a small number of CA containers are used for long-distance shipping of highly perishable, high-value products in the international trade. Refrigerated containers and trailers are more often used for longdistance shipping, whether by sea, rail or truck.

Shipping by refrigerated trucks is not only convenient, but also effective in preserving the quality of product. However, both the initial investment and the operating costs are very high. Another possibility is insulated or properly ventilated trailer trucks. Precooled products can be transported by well insulated non-refrigerated trucks for up to several hours without any significant rise in product temperature.

There are considerable cost savings without any sacrifice of quality if trucks are only insulated, rather than refrigerated, for short-distance shipping. If the product is not precooled or if the shipping distance is long, a ventilated truck is a better choice than an insulated truck without ventilation and without refrigeration. Ventilation alone does not usually provide a uniform cool temperature, but it may help dissipate excessive field heat and respiration heat, and thus avoid high temperature injury.

Temperature management is critical during long distance transport, so loads must be stacked to enable proper air circulation to carry away heat from the produce itself as well as incoming heat from the atmosphere and off the road. Transport vehicles should be well insulated to maintain cool environments for pre-cooled commodities and well ventilated to allow air movement through the produce.

During transport, produce must be stacked in ways that minimise damage, then be braced and secured. An open air vehicle can be loaded in such a way that air can pass through the load, and provide some cooling of the produce as the vehicle moves. Traveling during the night and early morning can reduce the heat load on a vehicle that is transporting produce.

Mixed loads can be a serious concern when temperature optima are not compatible (for example, when transporting chilling sensitive fruits with commodities that require very low temperatures) or when ethylene producing commodities and ethylene sensitive commodities are transported together. High ethylene producers (such as ripe bananas, apples, cantaloupe) can induce physiological disorders and/or undesirable changes in color, flavor and texture in ethylene sensitive commodities (such as lettuce, cucumbers, carrots, potatoes, sweet potatoes).

For domestic transportation the use of road vehicles offers substantial advantages of convenience, availability, flexibility permitting door-to-door delivery, and reasonable cost of transport. The use of road transportation for fresh produce is increasing and likely to increase in countries all over the world. Produce may be transported by pick-up, enclosed truck, open truck or refrigerated vehicle.

Enclosed Vehicle Transportation

Enclosed vehicle these re only suitable for short journeys, unless provided with a cooling system, since the produce inside heats up rapidly. However, they protect the produce from pilferage and physical injury, and are often used for urban retail delivery.

Open Vehicle Transportation

Pick-ups and open trucks are the commonest type of road transport. They are often fitted, before or after purchase, with frames to ease stacking and covering. Natural ventilation is usually sufficient to prevent overheating of produce over relatively short journeys and the most versatile types have a fixed roof and tarpaulin drapes which can be pulled along the sides and back to allow access for loading and unloading at any point. These loose awnings are not in contact with the produce and so allow for ventilation systems on these types of vehicles are unnecessary for short journeys, but when transit time are more than a few hours, adjustable louvres and air-scoops may be needed.

Bulk loads

Bulk loads of produce should be carefully loaded so as not to cause mechanical damage. Vehicles can be padded or lined with a thick layer of straw. Woven mats or sacks can be used in the beds of small vehicles. Other loads should not be placed on top of the bulk commodity.

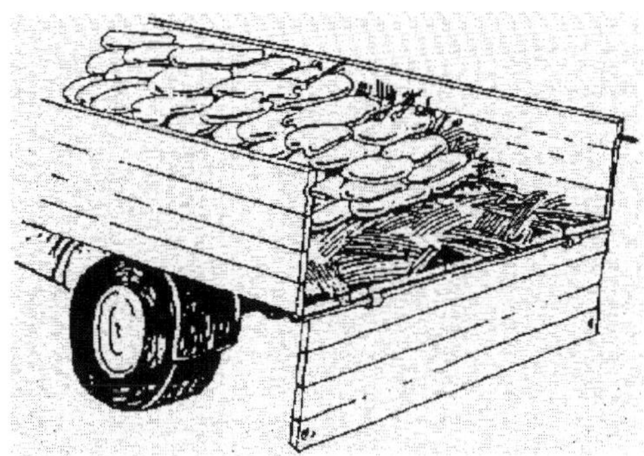

Bulk loads

Cooling open loads is desirable whenever possible. A truck ventilating device can be constructed for an unrefrigerated open vehicle by covering the load loosely with canvas and fashioning a wind catcher from sheet metal. Scoop should be mounted at the front of the bed and should reach somewhat higher the height of the cab.

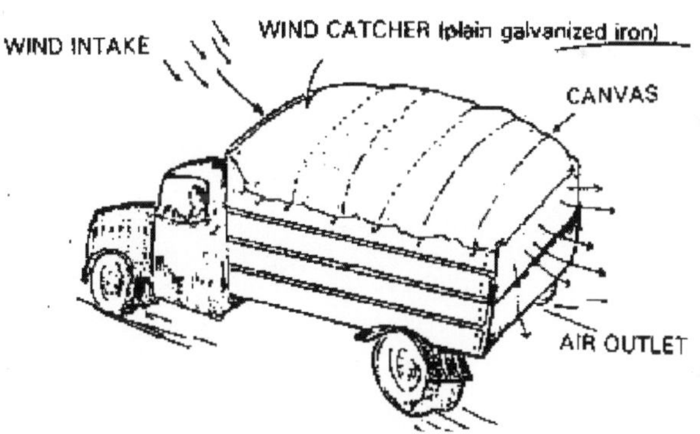

Cooling open loads

High transportation speeds and/or long distance transport run the risk of causing excess drying of the crop. This ventilating system was designed for hauling bulk loads of fresh fava beans in Iran. The wind catcher and ducts were constructed using wooden crates. After removing their end panels the crates were wired together into the pattern shown below.

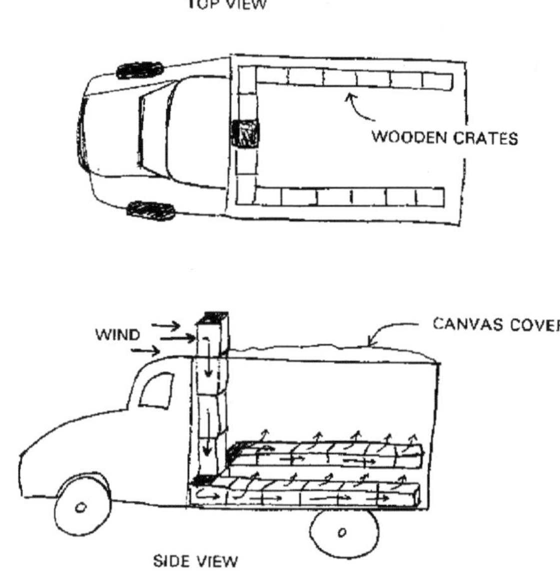

Open vehicle ventilating system

Air flows upward through the load during transport, helping to keep the produce from overheating. This system has also been used in pick-up trucks, and for hauling bulk greens and green beans. Best results are obtained when transporting during the early morning hours, before sunrise.

Refrigerated Vehicle Transportation

Some highly perishable products may justify the use of refrigerated vehicles. Ice is not generally used to refrigerated trucks because of weight and corrosion constraints and for most refrigerated vehicles a mechanical system is used. Truck-based mechanical refrigeration systems vary in their cooling capacity. Most are only suitable for maintaining the temperature of produce which has been pre-cooled by other means, and they possess fairly weak air-circulation fans sufficient only to allow refrigeration of air heated up by low respiration of the cool produce. Some form of ventilation may be necessary on longer journeys to prevent oxygen depletion and carbon dioxide accumulation. A few refrigerated vehicle types, usually reefer containers mounted on the back of flatbed trucks, are capable of rapid forced-draught cooling of warm produce, but these are generally an exception because of their expense.

There is often a tendency to use the relatively low capacity refrigerated trucks as the precooling system for export produce. The trucks were not designed as pre-coolers and the results are not satisfactory. In other instances, these same trucks may be used as mobile refrigerators to be installed for days or weeks at a time, as a form of refrigerated store. Again the results are far from satisfactory and spoilage levels are very high. In addition, use of the vehicles in this way is a waste of a very expensive transport vehicle.

Stacking Patterns

Handstacked

In order to reduce the conduction of heat from outside the vehicle into the load of produce, stacking patterns should reduce the amount of contact between the produce and the floor and walls of the vehicle.

Illustrated below is pyramid stacking of bags inside a refrigerated trailer.

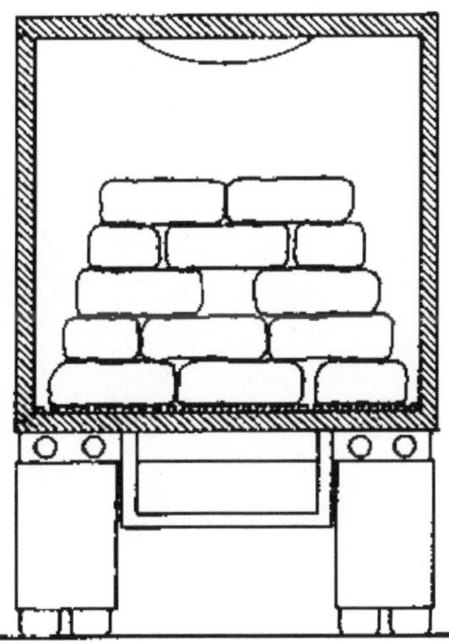

Pyramid stacking of bags inside a refrigerated trailer

Bushels of produce can be loaded into a refrigerated trailer using a pattern of alternately inverted layers that leave plenty of space between rows for air circulation.

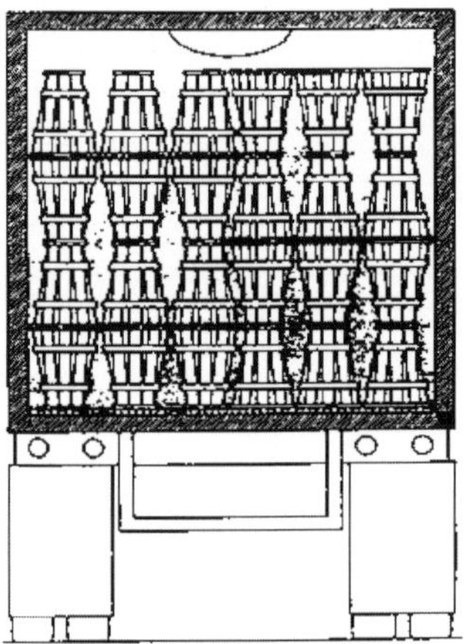

Pattern of alternately inverted layers

Produce transported in cartons should also be stacked so as to allow adequate air circulation throughout the load. The diagram below illustrates cross-wise offset loading of partial telescopic containers. On the floor of the truck, pallets or other supports should be used to keep the cartons out of direct contact with the floor.

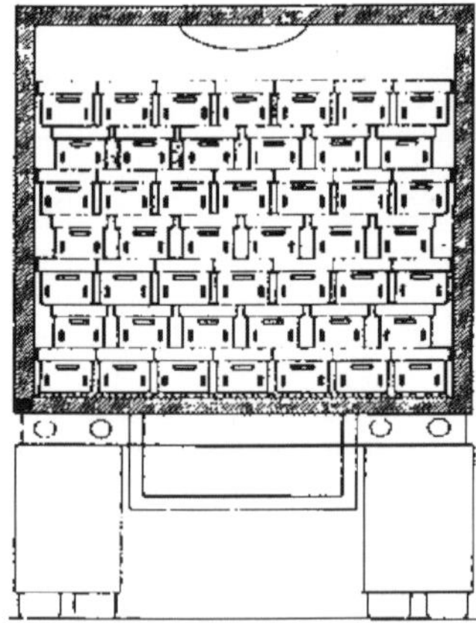

Cross-wise offset loading of partial telescopic containers

When cartons of various sizes must be loaded together, the larger, heavier containers should be placed on the bottom of the load. Parallel channels should be left for air to move through the length of the load.

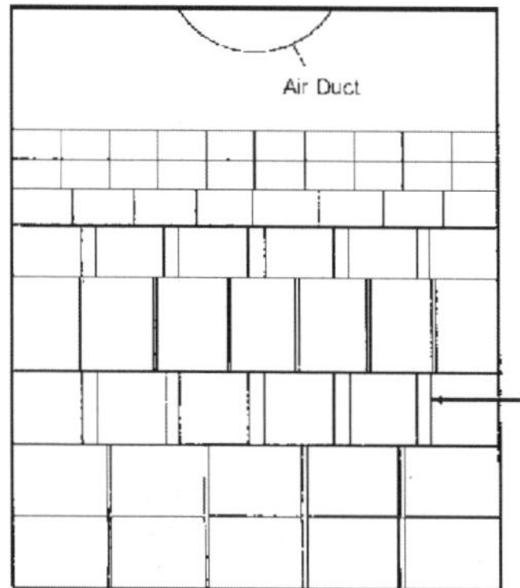

Parallel channels for air passage

Often the large containers used for cut flower packaging must be handstacked when loaded into a transport vehicle. The best loading pattern for cut flowers is known as the pigeon hole pattern, where boxes are stacked in alternating solid and open layers, and channels are left down both side walls.

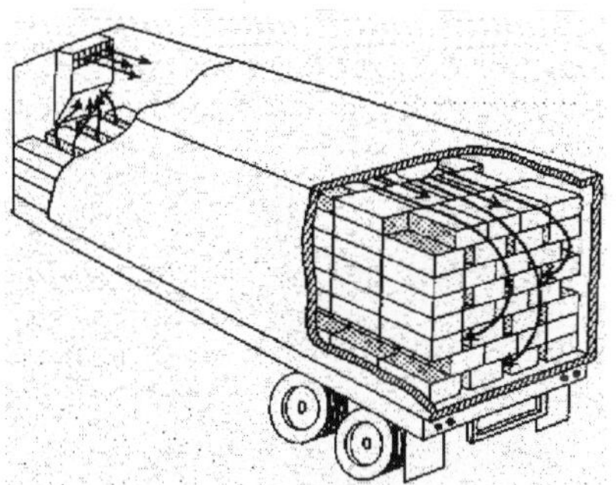

Pigeon hole pattern

This pattern provides channels for air circulation lengthwise through the load, and allows every box to be in direct contact with refrigerated air.

Pallet and Slip Sheet Loads

Containers should be loaded so that they are away from the side walls and the floor of the transport vehicle in order to minimise the conduction of heat from the outside environment. In the diagrams below, the numbers of cartons refer to how many cartons would be in contact with the walls and floor of the truck when fully loaded. Only the load on the bottom right is fully protected from heat transfer. The use of pallets keeps the cartons off the floor, while center-loading leaves an insulating air space between the pallet loads and the outside walls.

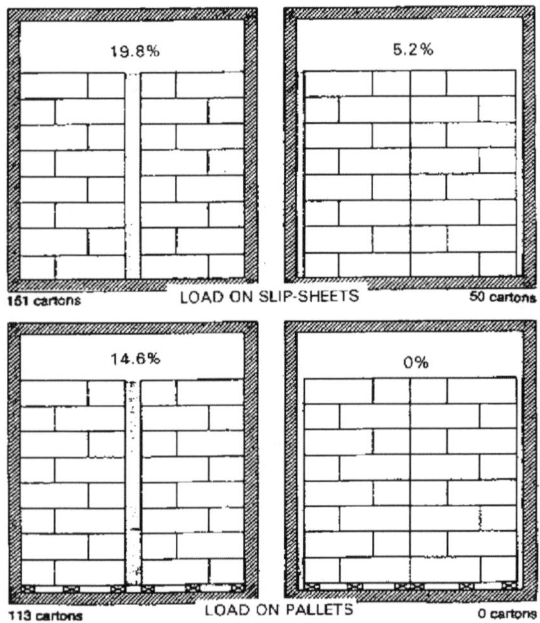

Pallet loads and slip sheets

Bracing the Load

There should always be a void between the last stack of produce and the back of the transport vehicle. The load should be braced to prevent shifting against the rear door during transit. If the load shifts, it can block air circulation, and fallen cartons can present great danger to workers who open the door at a destination market. A simple wooden brace can be constructed and installed to prevent damage during transport.

Sea Transport

The inter-island trade in fresh produce utilises a range of small vessels of wooden or steel construction which are independently owned and operated and focus almost exclusively on servicing the transport needs of the trade. The vessels all rely on engines for propulsion, but the wooden sloops regularly use their sails to stabilise and propell them. The wooden sloops are generally newer (ten years old or less) and carry only about 35 to 40 tonnes of cargo. The steel vessels by comparison are much older and most were bought cheaply and fairly recently second-hand from Europe where they were considered too old or expensive for continued service. Many of the steel vessels are 50 years or older and carry 60 to 100 or more tonnes of cargo.

Some of the larger steel vessels, such as the 'Stella SII' and the 'Louise Kingcraft', carry other types of cargo on a regular or opportunistic schedule between their regular fortnightly huckster route. Several of the small wooden sloops operating out of Dominica also double as fishing boats - fresh produce from Dominica is off-loaded in Antigua then the vessel will fish the waters in the Leewards and sell their catch at the weekend market in Antigua before returning to Dominica for more fresh produce.

All of the rest of the inter-island vessels concentrate solely on the fresh produce trade, but it should be noted that the return journey with dry goods and other consumables ranging from snack foods from Trinidad or cylinders of LPG gas and spare parts etc., is equally important to both the hucksters and traffickers in terms of cargo volume and probably more important in many cases in terms of profits.

Early in 1987, a Consultant Marine Engineer with the FAO Inter-Island Trade Project conducted a survey of, most of the vessels involved in the inter-island trade in fresh produce at that time. His conclusions may be summarised as follows:

— *General condition* - The condition of most of the vessels themselves was found to be reasonably good. Only the 'Bigdin' was found totally unsatisfactory, but then it was built in 1898 and should have been scrapped a long time ago. Many of the vessels regularly experienced engine breakdowns for lack of parts or occasionally lack of maintenance.

— *Fitness for fresh produce* - Not one single vessel of either steel or wooden construction was found to be appropriate for fresh produce transportation because of:

  — *Poor ventilation* - none of the vessels had adequate ventilation of the hold and most had no ventilation at all;

  — *No insulation of the engine compartment* - in most of the wooden vessels the engine compartment is not separated from the cargo hold and heat and fumes from the engine mixes with the cargo, often the engine sucks air from the cargo hold because there is no other air supply for engine operation;

— *No cargo liftinb equipment* - few of the vessels had equipment to assist with loading/unloading of cargo and much damage was impacted on the produce from rough handling by the crew and stevedores;

— *Cargohold shape* - Only the larger steel vessels offered a practical shape for loading of crates and boxes and the wooden vessels shape meant that sacks and bags were crammed into the spaces between and under heavy wooden crates.

The FAO Consultant recommended various inexpensive and practical improvements for the vessels and these were presented to the captains and owners for their guidance. The FAO project offered to financially assist with fitting of one or two demonstration vessels with the prescribed improvements, but met with no response and to date none of the vessels has been improved at the owner's or captain's initiative. The improvements recommended were:

— *Improved ventilation*—by fitting "cowl-type" ventilators as shown below.

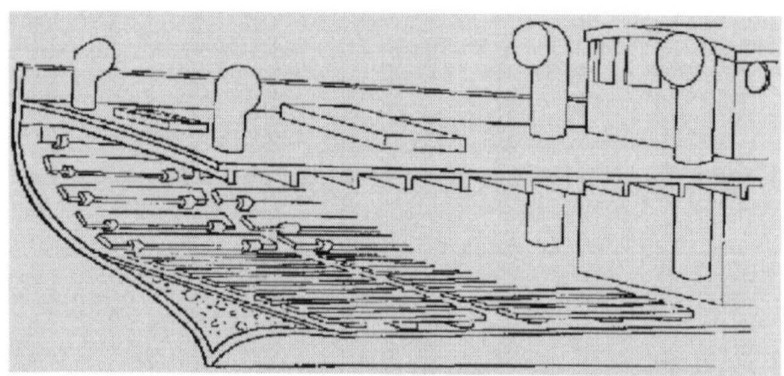

"Cowl-type" ventilators for inter-island vessels

Most of the vessels are not equipped with generators to provide power for electric fans and blowers, so natural airflow and some wind assistance can be used to direct air into the bottom of the cargo and engine holds separately and extract the warm air from the top of the holds. Cost is estimated at approximately US$2, 500 for each vessel assuming only one cargo hold.

— *Insulation of engine compartment* - by construction of a proper barrier between the engine and the cargo hold and fitting an insulating layer to prevent heat from the engine entering the cargo hold. The insulation could be prefabricated polystyrene panels, or steel coated polystyrene sandwich panels, or polyester foam insulation sprayed into place or onto spercially prepared plywood panels. All costs include wood, plywood and other materials as well as the insulation material itself.

— *Cargo lifting equipment* - intended only for the wooden sloops. Figure shown below is a suitable installation which can be fitted to the mast and yet would not interfere with normal sailing operations. At a later stage owners could fit a small diesel powered winch to further improve the design.

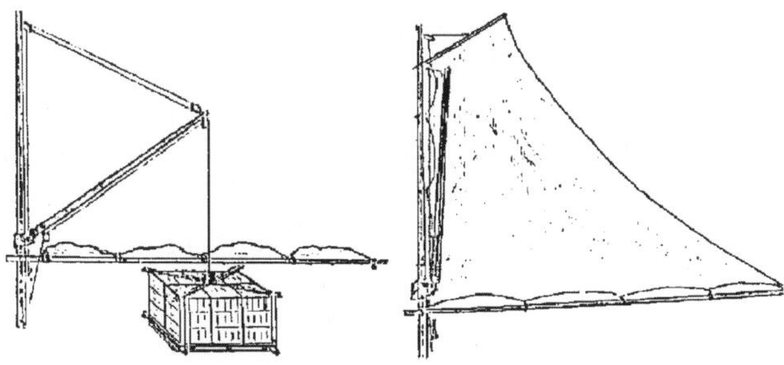

Cargo lifting equipment for inter-island schooner

Very few of the inter-island vessels carry any form of insurance because the owners claim the premiums are too high. The history of the inter-island trade is characterised by many sinking of wooden vessels with a complete loss of cargo and occasionally the lives of the hucksters accompanying their produce. Today, only a few of the steel vessels carry passengers and most of the traders fly to their destinations and meet the cargo there. The exceptions are still the very small open boats serving Marie Galante and Les Saintes from Dominica which still carry passengers and still experience losses of life and cargo.

Without insurance of their vessels, the owners cannot obtain loans for improvements to the vessels. However, the profits from the trade have been estimated on several occasions to be easily sufficient to support the cost of the above recommended improvements from cashflow alone and loans should not be necessary for any but the most poorly operated vessels. It would appear that the principal barrier to improving the vessels is the intransigence of the captains and owners content to reap their profits without the bother of improving the trade. Hucksters and traffickers generally lack sufficient organisation and cooperative spirit to force the owners into making improvements, and some owners are hucksters themselves.

Various reports, recommendations and proposals have been made by different aid organisations in the region to introduce an independent freight service based on modern and possibly purpose built vessels more suited to the transport of fresh produce. To-date, none of these suggested interventions has become a reality, and the inter-island trade will continue to suffer high postharvest losses of fresh produce because of inadequate transport service and conditions. The dilemma is such that some improvements can be made by introducing better forms of packaging but these will be limited until improvements are made in the vessels themselves. In addition, most of the vessels are really only suited to the use of sacks and bags because of the size and shape of the cargo hold.

Refrigerated Sea Transport

The perishability of fresh produce, allied with its tendency to heat up in confined spaces leading to rapid spoilage and decay, are all reasons why long distance unrefrigerated ship transportation is seldom used and never without high levels of spoilage. It is not likely that any mayor advances will be made in unrefrigerated shipping design to make the transport of fresh produce less risky. In most circumstances sea transportation is by reefer vessel and is largely used for export of fresh produce. Sea transportation, because of the journey times, is effectively a form of refrigerated storage and all the precautions necessary for storage are relevant here also.

Reefer vessels

Reefer vessels are totally refrigerated, have efficient air circulation systems and controllable rates of air-exchange. Loading is facilitated by side-hatches or by specialised loop-belt continuous conveyors, which transport the individual packages from the loading wharf up above the central hatches of the vessel and down into the holds (they are used in an identical fashion for off-loading). Reefer vessels are generally high capacity (4000 tones or more) and regularly carry fresh produce all over the world. The limiting factors are the journey times, which may exceed the storage-life of most produce, and the considerable amount of handling involved in loading and unloading.

Palletisation of produce has reduced much of the handling but break-bulk loading of individual packages is still widely used.

Reefer vessel transportation usually involves the export of large volumes of fresh produce and necessitates the dedicated activities of full-time personnel employed by a large volume producer. For example in the Eastern Caribbean, the banana operations. Chartering, if to be successful means organising regular large volume supplies of produce over an extended period or charter vessels will not be available.

Reefer containers

Reefer containers are a specialised form of sea transportation which is rapidly growing in international popularity. Each container either has its own independent refrigeration system which can be powered electrically by the container vessel, or has special air ducts are one end which are lined up with conduits on the container vessel and refrigeration is thus provided entirely by the vessel's own system.

Reefer containers are in standard sizes all of which are 8 x 8 ft. cross section, but which may be 10, 20, 30 or 40 ft. long. The most widely used sizes are the 40ft. followed by the 20ft. size. The container can be bought or leased but they can be very expensive and can vary greatly in quality and performance. Their principal advantages are:

— permit shared use of a reef container vessel by many producers of different commodities, provided they have access to containers and are exporting by the same route.

— greatly reduce handling damage since they are loaded at the packhouse and may not be unloaded until they reach a customer store in the country of export.

— temperature control is independently set and monitored.

— capable of rapid pre-cooling of produce from ambient tropical conditions.

However, their disadvantages are:

— very expensive to own or lease.

— large and heavy and require special lifting equipment.

— need to lease more containers than are in actual fuse because of delayed return and breakdown.

— not all countries have container handling facilities which may limit loading ports and discharge ports, although some vessels are self-loading and unloading.

Air Transportation

Air transportation is very expensive and usually can only be justified for high value export produce such as exotic tropical fruits and vegetables for the extra-regional markets. These markets are very sophisticated and demand top quality produce which is carefully packed in standardised fibre-board cartons and correctly labelled. Any produce not meeting these specifications, or of less than top quality, will either be rejected immediately or will be down-graded to a price level which gives a break-even price for the exporter or very often a loss on the consignment.

All air-freighted exports require a high degree of market research, planning, organisation and management. Constant communication with identified importers is vital to guage market trend, prices and fluctuations in demand, together with feedback on quality control.

Costs of air freight will vary according to distance carried and whether a charter or scheduled flight is used. If production and package costs are added it can very quickly be seen that profit margins are low. Some countries see export of these commodities less as a profit maker and more as an earner of valuable foreign exchange. Lack of planning and good management will result in poor quality, rejection and possible loss of income.

Airport Handling

Airports are designed to keep people away from cargo areas, but it is essential that personnel are given clearance for supervising the loading of produce onto the plane. The produce must be at the airport well before the time of flight departure. Delays are common and so facilities for holding the produce at the airport are necessary, and they must either be refrigerated or at the very least shaded. The loading supervisor or another person should have all necessary paperwork processed in good time and details of the consignment should be sent to the importer by telex or telefax.

Air Containers

Fresh produce may be sent by cargo carrier but more usually in the cargo hold of a passenger aircraft. The quantity that can be sent will vary according to aircraft and space available. Most aircraft servicing the Eastern Caribbean employ unit load containers (ULD), which fit the contours of the hold, as well as carrying goods packed loose. Others may use special thin aluminium pallets which are moved over rollers. These can be leased and a supply kept at the packhouse with the necessary securing nets. By palletising in this way at the packhouse and perhaps using refrigerated trucks the handling of the produce is greatly reduced and loading times for the aircraft become very rapid.

Temperature and Pressure

Cargo holds are frequently maintained at the same temperature and pressure as passenger areas, but this may not always be the case. Very often chill temperatures at high altitude can cause irreversible damage to the produce and flight conditions should always be checked with the airline. Of the aircraft currently servicing the Eastern Caribbean the situation tends to be confusing in that some holds may be temperature-controlled but at preset temperatures, and no choice is offered as to the temperature set. Other holds on the same aeroplane may have no temperature control and operate at ambient. Exporters must therefore select which produce must go in which hold well ahead of time and book that space with the airline if chilling of the produce is to be avoided.

Trans-shipment

Trans-shipment of fresh produce from one flight to another at an intermediate country airport is nearly always inviting disaster unless personnel are available to supervise. For example, air transport from Dominica and St. Vincent with only small air-fields must be trans-shipped out of the other islands airports which can accommodate large jets and have regular scheduled services to the extra-regional markets. The exporters must then rely upon the others to make sure that the trans-shipment occurs smoothly.

References

Kitinoja, L. and J.R. Gorny. 1999. *Post-harvest technology for small-scale produce marketers: Economic opportunities, quality and food safety.* Davis: University of California, Postharvest Horticulture Series 21.

Thompson, J.F., P.E. Brecht, R.T. Hinsch, and A.A. Kader, 2000. M*arine container transport of chilled perishable produce.* Oakland: Univ. Calif. Div. Agric. and Nat. Res. Publ. 21595, 32 pp.

9

Handling at Destination

When handling produce at its destination, again it is important to avoid rough handling, minimize the number of handling steps, and maintain the lowest feasible temperature. If produce is to be stored before sale, then wholesale and retail markets need clean, well insulated storage rooms. Since a variety of commodities is usually being handled simultaneously at this point, it is important to remember not to mix those with different temperature requirements or store ethylene sensitive commodities near ethylene generating commodities. Stacking of non-uniform containers should be done with care to prevent collapse of weaker packages.

Before produce is sold to the consumer, the handler may wish to sort for quality, or at least to discard any damaged or decayed produce in order to give the product more market appeal. If ripeness or maturity is non-uniform, sorting at destination can provide the seller with a higher price for the better quality produce.

If the produce handled is a climacteric fruit crop that was harvested before it was ripe, the handler at destination may want to ripen the produce before it is sold to the public. Sometimes commodities such as bananas are left at ambient temperatures and allowed to ripen naturally. Covering the bananas with a plastic sheet will help ripening be more uniform throughout the lot. Placing a simple air vent into the center of the pile of ripening fruit can help reduce overheating during ripening and increase subsequent shelf life. The introduction of ethylene gas or ethylene-releasing compounds into a special storage environment is a more effective way to ensure uniform ripening.When displaying horticultural crops, single or double layers of produce are most likely to protect the commodities from compression damage and overhandling by the consumers. Temperatures of the display tables or refrigerated supermarket displays should be suited to the commodity on sale. For example, while peppers and tomatoes look pleasing when displayed with lettuce, peppers and tomatoes are chilling sensitive, while lettuce is not. Misting commodities that can tolerate surface water with coot clean water can help maintain a high relative humidity around the product.

Outdoor marketplaces suffer from a lack of temperature control and high air circulation, which can lead to dessication of crops, which will be seen as shriveling and wilting. These marketplaces can often benefit by the increased use of shading and protection from prevailing winds.

Finally, the handler at destination can help reduce losses in the future by maintaining good records of the sources of losses suffered at the wholesale or retail level. Identifying whether losses were due to mechanical damage, decay/disease, immaturity or overripeness allows the handler to provide better quality feedback to produce suppliers.

Unloading

A loading dock can ease the work associated with handling horticultural produce at destination. Containers can be transfered more rapidly and with less bending and lifting. For large trucks, a loading dock 117-122 cm high functions well, while for small trucks or pickups a height of 66-81 cm is recommended.

A simple device for easing the work of unloading transport vehicles can be constructed using a two pulleys and a strong rope. One pulley can be mounted inside the truck at the front of the bed, the second outside on a portable post or on a stationary object like the side of a building. Containers can be hung directly by their handles or placed into a sling.

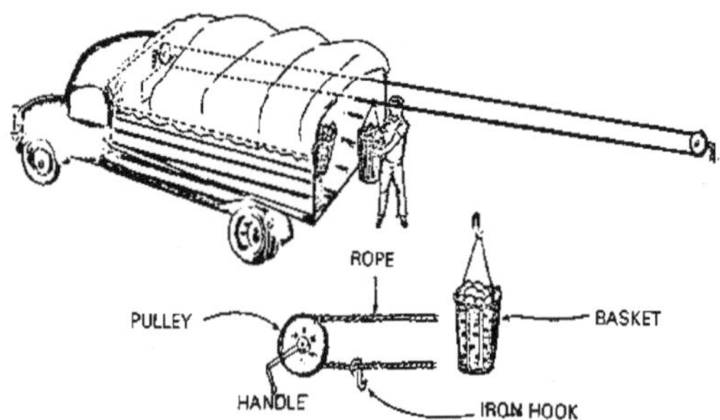

A simple set of stairs can be constructed to ease the work of loading and unloading produce. The stairs illustrated here can be folded and pushed back under the back of the truck when the vehicle is in motion. The steps can be made of wood or steel matting and steel bars can be used for supports.

Using a ramp is a simple and safe method for unloading produce. The ramp should be wide enough to prevent accidents and strong enough to carry the full weight of the handler plus the package she/he is carrying. Providing handtrucks or small carts can also ease the work associated with unloading.

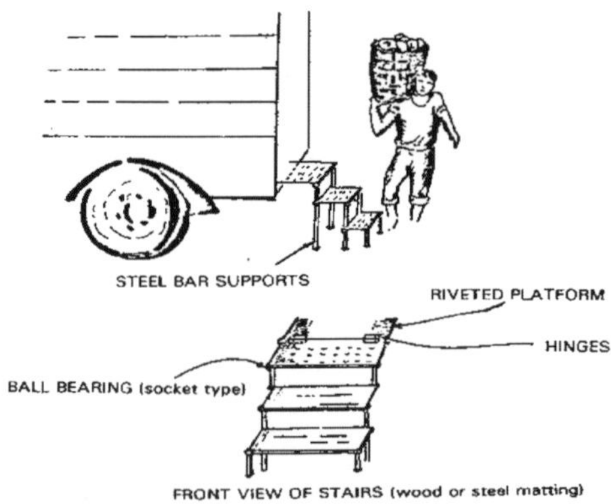

Storage temperatures

When produce is held at destination for a short time before marketing, the handler can help maintain quality and reduce losses by storing commodities at their most suitable temperature. However, if the storage period is five days or less, relative humidity is maintained between 85 and 95%, and the ethylene level is kept below 1 ppm ventilating or using a scrubber, most commodities can be grouped into the following three categories.

Table 1. Storage temperatures

32-36°F, 0-2°C			*45-50°F, 7-10°C*	*60-65°F, 16-18°C*
Vegetables and Melons				
anise	collard	parsnip	basil	casaba melon
artichoke	cut vegetables	raddichio	beans; snap, etc.	cassava
arugula	daikon	radish	cactus leaves	crenshaw melon
asparagus	endive	rutabaga	cucumber	dry onions
bean sprouts	escarole	rhubarb	eggplant	ginger
beet	garlic	salsify	Juan Canary	honeydew melon
Belgian endive	green onion	shallot	melon	jicama
bok choy	herbs(not basil)	spinach	kiwano	potato
broccoli	horseradish	snow pea	okra	Persian melon
broccoflower	Jerusalem	sweet corn	pepper; bell,	pumpkin
brussel sprouts	artichoke	sweet pea	chili	squash; winter,
cabbage	kale	Swiss chard	squash; summer,	hard rind
cantaloupe	kohlrabi	turnip	rind	sweet potato
carrot	leek	turnip greens	tomatillo	taro
cauliflower	lettuce	waterchestnut	watermelon	tomato; ripe
celeriac	mint	watercress		mature green
celery	mushroom			yam
chard	mustard greens			
chicory	parsley			
Fruits				
apple	quince		avocado, unripe	atemoya
apricot	raspberry		cactus pear, tuna	banana
avocado, ripe	strawberry		carambola	breadfruit
blackberry			chayote	cherimoya
blueberry			cranberry	coconut
cherry			feijoa	grapefruit
currant			guava	lemon
cut fruits			kumquat	lime
date			longan	mango
fig			lychee	mangosteen
gooseberry			mandarin	papaya
grape			olive	plantain
kiwifruit			orange	pummelo
nectarine			passion fruit	rambutan
peach			pepino	sapote
pear; Asian			pineapple	soursop
European			pomegranate	
persimmon			tamarillo	
plum			tangelo	
prune			tangerine	

Sorting/repacking

Some produce may require washing, trimming, bunching or sorting at the wholesale or retail market level. The layout of the work station used for handling produce at destination should be organized to minimize non-productive movement. In the illustration below, a dump table is located next to a sink for washing produce, and the drain board is positioned directly next to the sink. Once produce has dried, cartons can be packed and placed onto a cart located right next to the repacking table. With this layout, a single operator could easily perform all the handling steps or several handlers could work side by side.

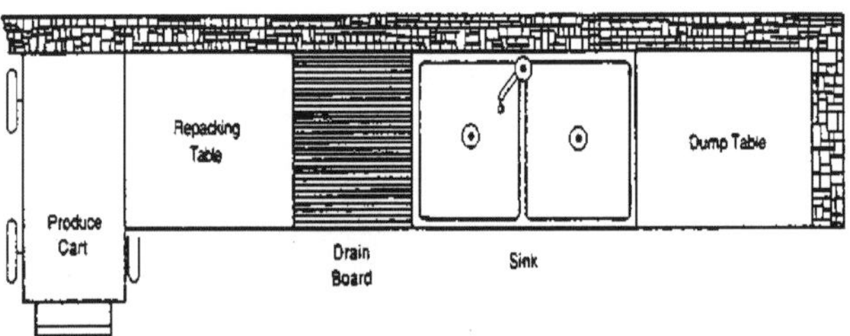

Some produce may have to be repacked by the wholesaler or retailer due to changes in quality or uneven ripening. The tomato sorting table illustrated below has work stations for up to five who select either ripes, pinks or breakers and allow the green tomatoes to run off to the end of the line. Rejects (culls) are placed in pails under the table.

Ripening

Some commodities may require ripening before sale at the wholesale or retail level. Ripening rooms are often used for tomatoes, citrus fruits and bananas. The use of diluted ethylene gas mixtures is safer than using pure ethylene which is explosive and flammable at concentrations of 3% or higher.

For tomatoes, technical grade ethylene gas is introduced into the room at a concentration of about 100 ppm for about 48 hours. Approximately 0.25 cubic feet/hr of ethylene gas is required for each 1000 cubic feet of ripening room volume. A small fan can be used to ensure a uniform continuous flow of ethylene into and through the room. Forced-air ripening is increasingly being used to provide more uniform temperatures and ethylene concentrations throughout the ripening room.

Ripening can also be initiated by using ethylene generated by passing ethanol over a bed of activated alumina. This method is safer than using pure ethylene gas. Ethylene-releasing compounds such as ethephon {(2-chloroethyl) phosphoric acid} are sometimes used to ripen tomatoes destined for processing. When using ethephon as a spray, the amount of ethylene released will increase as pH and/or relative humidity increase.

Several small ripening rooms may be more useful than a single large room for small scale handlers, since the amount of product handled at destination may vary from time to time. In this case, flow through systems can be designed to allow the use of one or more rooms at the same time.

Flowmeters can be located in one place for ease of monitoring, or can be strung out in a line. Locating all the flowmeters in one place requires the use of more tubing than if flowmeters are located in each ripening room. For more information and details on how to set up a flow through system for ripening fruit.

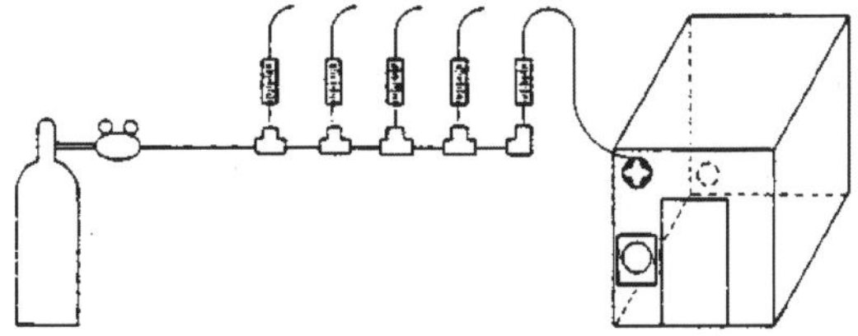

Flowmeters in one location

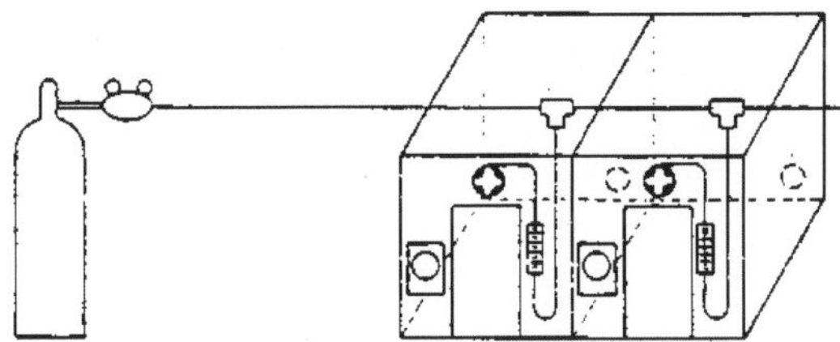

Flowmeters located in each ripening room

Small-scale handlers can now lease portable ripening facilities from a variety of companies in the United States. The self-contained, portable system illustrated below features a 20 pallet capacity, high capacity air flow and is simple to operate. All that is needed is a loading dock and a supply of 220V electricity.

Small-scale wholesalers and retailers can ripen fruits in bins or large cartons by placing a small quantity of ethylene-generating produce such as ripe bananas in with the produce to be ripened. Cover the bin or carton with a plastic sheet for 24 hours, then remove the plastic cover.

A simple way to ripen fruits at home in small amounts is to use a ripening bowl. Fruits that require ripening should be placed into the bowl with a ripe apple or ripe banana (or any other high ethylene-generating product). The bowl shown below is made of clear molded plastic and has ventilation holes around the top. Using this method, ripening will take from one to four days.

Home ripening is also possible using another, extremely low-tech practice— place fruits to be ripened into a paper bag with a ripe piece of fruit, close loosely and check in a few days.

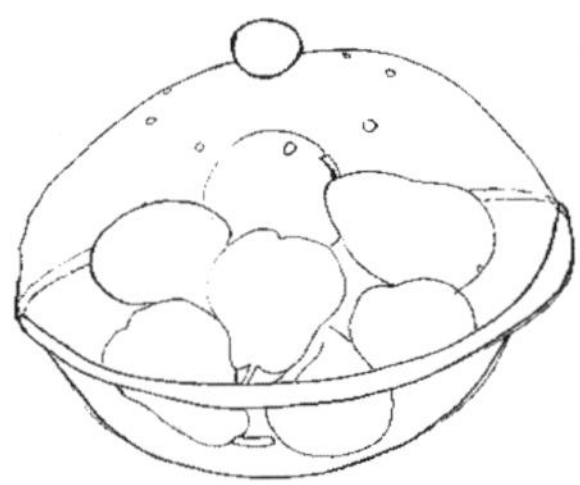

Display

This wooden display table given below is designed to be used for commodities such as cruciferous crops or leafy green vegetables that tolerate cooling with ice. The table can be used in the horizontal position or as a tilted display.

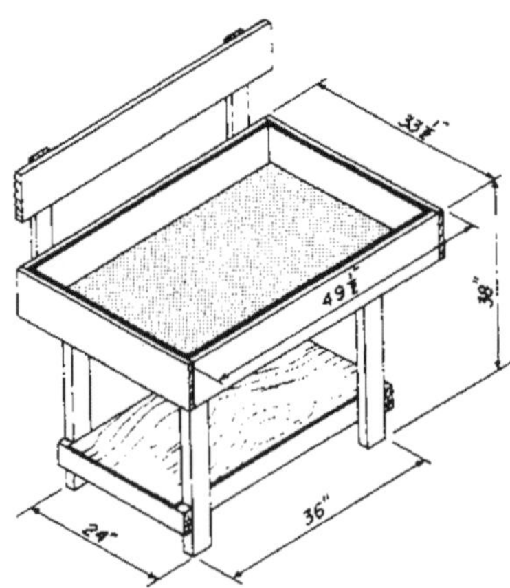

Four to five lbs of crushed ice per square foot of display space are required for cooling per day. A catch pail should be provided for melt water. To minimize ice needs, the display tray should be insulated and kept out of the direct sun.

High relative humidity can be maintained during display by misting leafy vegetables and water tolerant crops with clean, cold water. A simple sprinkler device can be constructed by perforating a pipe with tiny holes and connecting it to a hose. If this display is used outdoors, shade should be provided.

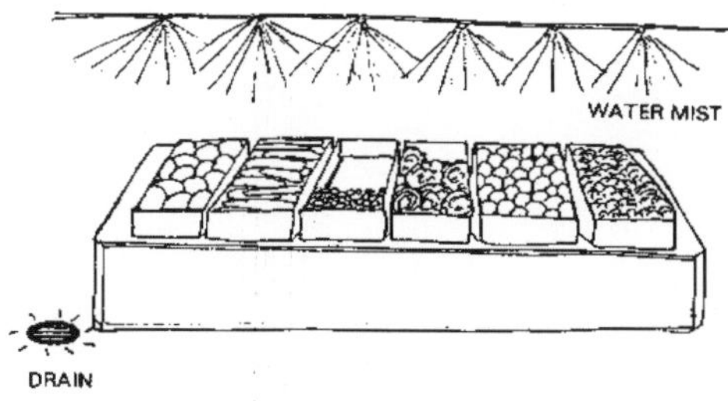

A simple semi-circular display table can be constructed from one four foot by eight foot sheet of plywood.

References

ASEAN-PHTRC. 1984. *Village Level Handling of Fruits and Vegetables: Traditional Practices and Technological Innovations.* Postharvest Horticulture and Training Center, College of Agriculture, University of the Philippines at Los Baños. Extension Bulletin No. 1

Kitinoja, L. and A.A. Kader. 2002. *Small-scale post-harvest handling practices: A manual for horticultural crops.* Fourth edition. Davis: University of California, Postharvest Horticulture Series 8E, 260 pp.

Seymour, G.B., J.E. Taylor, and G.A. Tucker, eds. 1993. *Biochemistry of fruit ripening.* London: Chapman and Hall, 454 pp.

Snowden, A.L. 1990. A color atlas of post-harvest diseases and disorders of fruits and vegetables. Vol. 1. *General introduction and fruits.* Boca Raton, FL: CRC Press, 302 pp.

10

Processing of Horticulture Crops

When conditions are not suitable for storage or immediate marketing of fresh produce, many horticultural crops can be processed using simple technologies. There are some processing methods that can be used by small-scale handlers, including drying, fermenting, canning, freezing, preserving and juicing. Fruits, vegetables and flowers can all be dried and stored for use or sale in the future. Fermentation is popular throughout the world as a food preservation method, and over 3,500 individual fermented foods have been described by Campbell-Platt. Fruits and vegetables can be canned or frozen, and fruits are often preserved in sugar or juiced.

The magnitude of post harvest losses in fresh fruits and vegetables is estimated to be 25-80 % depending upon the commodity and the technological level of post harvest operations. This reflects the lack of knowledge by post harvest handlers of the biological and environmental factors involved in the deterioration or the absence of adequate post harvest technologies required preserving fresh quality.

Concept of Water Activity (A_w) in Food Preservation

The concept of a_w has been very useful in food preservation and on that basis many processes could be successfully adapted and new products designed. Water has been called the universal solvent as it is a requirement for growth, metabolism, and support of many chemical reactions occurring in food products. Free water in fruit or vegetables is the water available for chemical reactions, to support microbial growth, and to act as a transporting medium for compounds.

In the bound state, water is not available to participate in these reactions as it is bound by water soluble compounds such as sugar, salt, gums, etc. (osmotic binding), and by the surface effect of the substrate (matrix binding). These water-binding effects reduce the vapour pressure of the food substrate according to Raoult's Law. Comparing this vapour pressure with that of pure water (at the same temperature) results in a ratio called water activity (a_w). Pure water has an a_w of 1, one molal solution of sugar - 0.98, and one molal solution of sodium chloride - 0.9669.

A saturated solution of sodium chloride has a water activity of 0.755. This same NaCl solution in a closed container will develop an equilibrium relative humidity (ERH) in a head space of 75.5%. A relationship therefore exists between ERH and a_w since both are based on vapour pressure. The ERH of a food product is defined as the relative humidity of the air surrounding the food at which the product neither gains nor loses its natural moisture and is in equilibrium with the environment.

The definition of moisture conditions in which pathogenic or spoilage microorganisms cannot grow is of paramount importance to food preservation. It is well known that each microorganism has a critical a_w below which growth cannot occur. For instance, pathogenic microorganisms cannot grow at aw <0.86; yeasts and moulds are more tolerant and usually no growth occurs at a_w <0.62. The so-called intermediate moisture foods (IMF) have a_w values in the range of 0.65-0.90.

Enzyme-catalysed reactions can occur in foods with relatively low water contents. The apparent cessation of the reaction at low moisture cannot be due to the irreversible inactivation of the enzyme, because upon humidification to a higher water activity, hydrolysis resumes at a rate characteristic of the newly attained water activity.

The relationship between water content and water activity is complex. An increase in a_w is usually accompanied by an increase in water content, but in a non-linear fashion. This relationship between water activity and moisture content at a given temperature is called the moisture sorption isotherm. These curves are determined experimentally and constitute the fingerprint of a food system. Many methods and instruments are available for laboratory measurement of water activity in foods. Methods are based on the colligative properties of solutions. Water activity can be estimated by measuring the following:

- — Vapour pressure
- — Osmotic pressure
- — Freezing point depression of a liquid
- — Equilibrium relative humidity of a liquid or solid
- — Boiling point elevation
- — Dew point and wet bulb depression
- — Suction potential, or by using the isopiestic method
- — Bithermal equilibrium
- — Electric hygrometers
- — Hair hygrometers

Vapour Pressure

Water activity is expressed as the ratio of the partial pressure of water in a food to the vapour pressure of pure water with the same temperature as the food. Thus, measuring the vapour pressure of water in a food system is the most direct measure of a_w. The food sample measured is allowed to equilibrate, and measurement is taken by using a manometer or transducer device as depicted in Figure below.

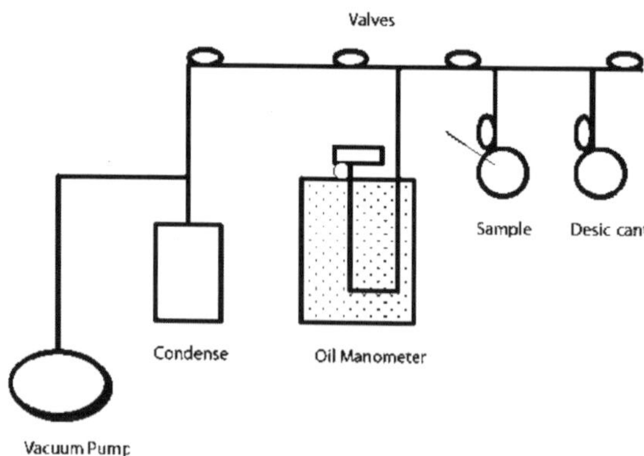

Vapour pressure manometer

This method can be affected by sample size, equilibration time, temperature, and volume. This method is not suitable for biological materials with active respiration or materials containing large amounts of volatiles.

Depression and Elevation of Freezing Point

This method is accurate for liquids in the high water activity range but is not suitable for solid foods. The water activity can be estimated using the following two expressions:

Freezing point depression:

$$-\log a_w = 0.004207\ DT_f + 2.1\ E\text{-}6\ DT^2_f \quad (1)$$

where DT_f is the depression in the freezing temperature of water.

Boiling point elevation:

$$-\log a_w = 0.01526\ DT_b - 4.862\ E\text{-}5\ DT^2_b \quad (2)$$

where DT_b is the elevation in the boiling temperature of water.

Osmotic Pressure

Water activity can be related to the osmotic pressure (p) of a solution with the following equation:

$$p = RT/V_w \ln(a_w) \quad (3)$$

where V_w is the molar volume of water in solution, R the universal gas constant, and T the absolute temperature. Osmotic pressure is defined as the mechanical pressure needed to prevent a net flow of solvent across a semi-permeable membrane. For an ideal solution, Equation (3) can be redefined as:

$$p = RT/V_w \ln(X_w) \quad (4)$$

where X_w is the molar fraction of water in the solution. For non-ideal solutions, the osmotic pressure expression can be rewritten as:

$$p = RTfnmb(m_w V_w) \quad (5)$$

where n is the number of moles of ions formed from one mole of electrolyte, m_w and mb are the molar concentrations of water and the solute, respectively, and f the osmotic coefficient, defined as:

$$f = -m_w \ln(a_w)/nmb \quad (6)$$

Dew Point Hygrometer

Vapour pressure can be determined from the dew point of an air-water mixture. The temperature at which the dew point occurs is determined by observing condensation on a smooth, cool surface such as a mirror. This temperature can be related to vapour pressure using a psychrometric chart.

Thermocouple Psychrometer

Water activity measurement is based on wet bulb temperature depression. A thermocouple is placed in the chamber where the sample is equilibrated. Water is then sprayed over the thermocouple before it is allowed to evaporate, causing a decrease in temperature. The drop in temperature is related to the rate of water evaporation from the surface of the thermocouple, which is a function of the relative humidity in equilibrium with the sample.

Isopiestic Method

The isopiestic method consists of equilibrating both a sample and a reference material in an evacuated desiccator until equilibrium is reached at 25°C. The moisture content of the reference material is then determined and the a_w obtained from the sorption isotherm. Since the sample was in equilibrium with the reference material, the a_w of both is the same.

Electric Hygrometers

Most hygrometers are electrical wires coated with hygroscopic salts or sulfonated polystyrene gel in which conductance or capacitance changes as the coating absorbs moisture from the sample. The major disadvantage of this type of hygrometer is the tendency of the hygroscopic salt to become contaminated with polar compounds, resulting in erroneous a_w determinations.

Hair Hygrometers

Hair hygrometers are based on the stretching of a fibre when exposed to high water activity. They are less sensitive than other instruments at lower levels of activity (<0.03 a_w) and the principal disadvantage of these types of meters is the time delay in reaching equilibrium and the tendency to hysteresis. Today we find many brands of water activity meters in the market. Most of these meters are based on the relationship between ERH and the food system, but differ in their internal components and configuration of software used.

One of the water activity meters most used today is the AcquaLab Series 3 Model TE, developed by Decagon Devices, which is based on the chilled-mirror dew point method. This instrument is a temperature controlled water activity meter that allows placement of the sample in a temperature stable environment without the use of an external water bath. The temperature can be selected on the screen and is monitored and controlled with thermoelectric components. Most of the older generations of water activity instruments are based on a temperature-controlled environment. Therefore, a margin of error greater than 5% can be expected due to temperature variations. This equipment is highly recommended for measuring water activity in fruits and vegetables since it measures a wide range of water activity.

The major advantages of the chilled-mirror dew point method are accuracy, speed, ease of use and precision. The AquaLab's range is from 0.030 to 1.000a_w, with a resolution of ±0.001a_w and accuracy of ±0.003a_w. Measurement time is typically less than five minutes. Capacitance sensors have the advantage of being inexpensive, but are not usually as accurate or as fast as the chilled-mirror dew point method. Capacitive instruments measure over the entire water activity range 0 to 1.00 a_w, with a resolution of ±0.005a_w and accuracy of ±0.015a_w. Some commercial instruments can complete measurements in five minutes while other electronic capacitive sensors usually require 30 to 90 minutes to reach equilibrium relative humidity conditions.

Intermediate Moisture Foods (IMF)

Traditional intermediate moisture foods (IMF) can be regarded as one of the oldest foods preserved by man. The mixing of ingredients to achieve a given a_w, that allowed safe storage while maintaining enough water for palatability, was only done, however, on an empirical basis. The work done by food scientists approximately three decades ago, in the search for convenient stable products through removal of water, resulted in the so-called modern intermediate moisture foods. These foods rely heavily on the addition of humectants and preservatives to prevent or reduce the growth of microorganisms. Since then, this category of products has been subjected to continuous revision and discussion.

Definitions of IMF in terms of a_w values and moisture content vary within wide limits (0.6-0.90 a_w, 10-50% moisture), and the addition of preservatives provides the margin of safety against spoilage organisms tolerant to low a_w. Of the food poisoning bacteria, Staphylococcus aureus is one of the organisms of high concern since it has been reported to tolerate a_w as low as 0.83-0.86 under aerobic conditions. Many of the considerations on the significance of microorganisms in IMF are made in terms of a_w limits for growth. However, microbial control in IMF does not only depend on a_w but on pH, Eh, F and T values preservatives, competitive microflora, etc., which also exert an important effect on colonising flora.

The application of IMF technology has been very successful in preserving fruits and vegetables without refrigeration in most Latin American countries. For instance, the addition of high amounts of sugar to fruits during processing will create a protective layer against microbial contamination after the heat process. The sugar acts as a water activity depressor limiting the capability of bacteria to grow in food.

Food products formulated under this concept are stable at room temperature without thermal processing and can be generally eaten without rehydration. Some processed fruits and vegetables are considered IMF foods. These include cabbage, carrots, horseradish, potatoes, strawberries, etc.; their water activities at 30°C follow:

Foods	a_w
Cabbage	0.64
	0.75
Carrots	0.64
	0.75
Horseradish	0.75
Potatoes	0.75
	0.64
Strawberries	0.65
	0.75

Under these conditions, bacterial growth is inhibited but some moulds and yeast may grow at a_w greater than 0.70. In addition, chemical preservatives are generally used to inhibit the growth of moulds and yeasts in fruits and vegetables.

Advantages of IMF preservation are: Intermediate moisture foods have an a_w range of 0.65-0.90, and thus water activity is their primary hurdle to achieving microbial stability and safety. IMF foods are easy to prepare and store without refrigeration. They are energy efficient and relatively cheap. They are not readily subject to spoilage, even if packages have been damaged prior to opening, as with thermostabilised foods, because of low a_w. This is a plus for many developing countries, especially those in tropical climates with inadequate infrastructure for processing and storage, and offers marketing advantages for consumers all over the world.

Disadvantages of IMF preservation are: Some IMF foods contain high levels of additives (i.e., nitrites sulphites, humectants, etc.) that may cause health concerns and possible legal problems. High sugar content is also a concern because of the high calorific intake. Therefore, efforts are been made to improve the quality of such foods by decreasing sugar and salt addition, as well as by increasing the moisture content and a_w, but without sacrificing the microbial stability and safety of products if stored without refrigeration. This may be achieved by an intelligent application of hurdles.

Fruit products from intermediate moisture foods (IMF) appear to have potential markets. However, application of this technology to produce stable products at ambient temperature is limited by the high concentration of solutes required to reduce water activities to safe levels. This usually affects the sensory properties of the food.

Combined Methods for Preservation

Food preserved by combined methods (hurdles) remains stable and safe even without refrigeration, and is high in sensory and nutritive value due to the gentle process applied. Hurdle technology is the term often applied when foods are preserved by a combination of processes. The hurdle includes temperature, water activity, redox potential, modified atmosphere, preservatives, etc. The concept is that for a given food the bacteria should not be able to "jump over" all of the hurdles present, and so should be inhibited.

If several hurdles are used simultaneously, a gentle preservation could be applied, which nevertheless secures stable and safe foods of high sensory and nutritional properties. This is because different hurdles in a food often have a synergistic (enhancing) or additive effect. For instance, modified foods may be designed to require no refrigeration and thus save energy. On the other hand, preservatives (e.g., nitrite in meats) could be partially replaced by certain hurdles (such as water activity) in a food. Moreover, a hurdle could be used without affecting the integrity of food pieces (e.g., fruits) or in the application of high pressure for the preservation of other foods (e.g., juices). Hurdle technology is applicable both in large and small food industries.

In general, hurdle technology is now widely used for food design in making new products according to the needs of processors and consumers. For instance, if energy preservation is the goal, then energy consumption hurdles such as refrigeration can be replaced by hurdles (a_w, pH, or Eh) that do not require energy and still ensure a stable and safe product. The hurdle effect is an illustration of the fact that in most foods several factors (hurdles) contribute to stability and safety. This hurdle effect is of fundamental importance for the preservation of food, since the hurdles in a stable product control microbial spoilage and food poisoning as well as undesirable fermentation.

Increasing consumer demand for fresh quality products is turning processors to the so-called minimally processed products (MP), an attempt to combine freshness with convenience to the point that even the traditional whole, fresh fruit or vegetable is being packaged and marketed in ways formerly reserved for processed products. According to these authors, the widely accepted concept of MP refrigerated fruits involves the idea of living respiring tissues. Because MP refrigerated products can be raw, the cells of the vegetative tissue may be alive and respiring (as in fruits and vegetables), and biochemical reactions can take place that lead to rapid senescence and/ or quality changes.

In these products, the primary spoilage mechanisms are microbial growth and physiological and biochemical changes, and in most cases, minimally processed foods are more perishable than the unprocessed raw materials from which they are made. The technology for shelf-stable high moisture fruit products (HMFP) is based on a combination of inhibiting factors to combat the deleterious effects of microorganisms in fruits, including additional factors to reduce major quality losses from reactions.

In order to select a combination of factors and levels, the type of microorganism and quality loss from reactions that might occur must be anticipated. Minimal processing may encompass pre-cut refrigerated fruits, peeled refrigerated whole fruits, sous vide dishes, which may include pre-heated vegetables and fruits, cloudy and clarified refrigerated juices, freshly squeezed juices, etc. All of these products have special packaging requirements coupled with refrigeration. These products, apart from special handling, preparation, and size reduction operations, might also require special distribution and utilisation operations such as Controlled atmosphere/ Modified atmosphere/air flow rate/vacuum storage (O_2, CO_2, N_2, CO, C_2H_2, H_2O controls), computer controlled warehousing, retailing and food service, communications network, etc. HMFP fruits are less sophisticated than MPR fruits and should be priced lower when introduced commercially.

Careful selection of these processes should of course be made to find the appropriate methods suited to a particular rural or village situation.

Water Activity Reducing Agents

Glucose: Glucose is not a very good humectant due to the lower water holding capacity (WHC), which makes it difficult to obtain the isotherm curve at low a_w.

Fructose: Fructose has a higher water activity reduction capacity and therefore is more desirable as a humectant in stabilising food products.

Sucrose: Sucrose is one of the most studied sugars and is widely used in food systems, in the confectionary industry, both in the U.S. and Europe, but has a lower water activity reduction capacity compared to fructose.

Other humectants: Based solely on the water activity reduction capacity, sorbitol and fructose are the most desirable humectants. Sucrose has the third best reduction capacity and lactose the poorest. The amorphous form absorbs more water at specific a_w than the corresponding crystalline form. NaCl and KCl salts appear to be superior humectants at a high range of a_w. The increased a_w lowering ability exhibited by the salts may be explained by the smaller molecular weight, increasing the ability to bind or structure more water.

Other sugars used as humectants in food stability include lactose and sorbitol. The amorphous form absorbs more water at specific a_w than the crystalline form. Polyols are better humectants than sugars because of their greater water activity reduction capacity and are less hygroscopic than sugars. The most widely used polyols as humectants in foods are 1,3- butyleneglycol, propylene glycol, glycerol, and polyethylene glycol 400.

pH Reducing Agents

Organic acids: Organic acids, whether naturally present in foods due to fermentation or intentionally added during processing, have been used for many years in food preservation. Some organic acids behave primarily as fungicides or fungistats, while others tend to be more effective at inhibiting bacterial growth. The mode of action of organic acids is related to the pH reduction of the substrate, acidification of internal components of cell membranes by ionisation of the undissociated acid molecule, or disruption of substrate transport by alteration of cell membrane permeability.

The undissociated portion of the acid molecule is primarily responsible for antimicrobial activity; therefore, effectiveness depends upon the dissociation constants (pKa) of the acid. Organic acids are generally more effective at low pH and high dissociation constants. The most commonly used organic acids in food preservation include: citric, succinic, malic, tartaric, benzoic, lactic, and propionic acids.

Citric acid is present in citrus fruits. It has been demonstrated that citric acid is more effective than acetic and lactic acids for inhibiting growth of thermophilic bacteria. Also, combinations of citric and ascorbic acids inhibit growth and toxin production of C. botulinum type B in vacuum-packed cooked potatoes.

Malic acid is widely found in fruits and vegetables. It inhibits the growth of yeasts and some bacteria due to a decrease in pH.

Tartaric acid is present in fruits such as grapes and pineapples. The antimicrobial activity of this acid is attributed to pH reduction.

Benzoic acid is the oldest and most commonly used preservative. It occurs naturally in cranberries, raspberries, plums, prunes, cinnamon, and cloves. As an additive, sodium salt in benzoic acid is suitable for foods and beverages with pH below 4.5. Benzoic acid is primarily used as an antifungal agent in fruit-based and fruit beverages, fruit products, bakery products, and margarine.

Lactic acid is not naturally present in foods; it is formed during fermentation of foods such as sauerkraut, pickles, olives, and some meats and cheeses by lactic acid bacteria. It has been reported that lactic acid inhibits the growth of spore forming bacteria at pH 5.0 but does not affect the growth of yeast and moulds.

Propionic acid occurs in foods by natural processing. It is found in Swiss cheese at concentrations up to 1%, produced by Propionicbacterium shermanii. The antimicrobial activity of propionic acid is primarily against moulds and bacteria.

Inorganic acids: Inorganic acids include hydrochloric, sulphuric, and phosphoric, the latter being the principal acid used in fruit and vegetable processing). They are mainly used as buffering agents, neutralisers, and cleaners.

Fermentation by-products: Fermentation by-products are formed during fermentation of fruits and vegetables, as in sauerkraut processing, pickling, and wine making. One by-product, lactic acid, is formed during fermentation of cabbage or cucumbers. This acid decreases the pH of fruits and vegetables, producing the characteristic flavour of sauerkraut, and acts as a controller of pathogens that may develop in the final fermented product.

Browning Inhibitors

Sulphites, bisulphites, and metabisulphites: Sodium bisulphite is a potential browning inhibitor in fruit and vegetable products (e.g., peeled potatoes and apples). This preservative when used in food production can delay or prevent undesirable changes in the colour, flavour, and texture of fresh fruits and vegetables, potatoes, drinks, wine, etc. Potassium bisulphite is used in a similar way to sodium bisulphite, and is used in the food industry to prevent browning reactions in fruit and vegetable products. Sulphites, bisulphites, and metabisulphites of both sodium and potassium together with gaseous sulphur dioxides are all chemically equivalent.

Sulphite levels in processed foods are expressed as SO_2 equivalents, and range from zero to about 3000 ppm in dry weight. Dehydrated, light coloured fruits (e.g., apples, apricots, bleached raisins, pears, and peaches) contain the greatest amounts in this range. Dehydrated vegetables and prepared soup mixes range from a few hundred to about 2000 ppm; instant potatoes contain approximately 400 ppm. The dose for wine is about 100-400 ppm and for beer about 2-8 ppm. The maximum legal sulphite level in wines permitted by the Food and Drug Administration (FDA) is 300 ppm. In the U.S. most wines have a sulphite level of 100 ppm.

Sulphites are highly effective in controlling browning in fruits and vegetables, but are subject to regulatory restrictions because of adverse effects on health. Sulphites inhibit non-enzymatic browning by reacting with carbonyl intermediates, thereby preventing further reaction. Sulphite levels in foods vary widely depending on the application. Residual levels never exceed several hundred per million but could reach 100 ppm in some fruits and vegetables. The maximum sulphur dioxide levels in fruit juices, dehydrated potatoes, and dried fruits permitted by the FDA are 300, 500, and 2000 ppm, respectively.

Additives

Potassium sorbate: Potassium sorbate is a white crystalline powder that has greater solubility in water than sorbic acid, which may be used accordingly in making concentrates for dipping, spraying, or metering fruit and vegetable products. It has antimycotic actions similar to sorbic acid, but usually 25% more potassium sorbate must be used than sorbic acid to secure the same protection. The common salt of potassium sorbate was developed because of its high solubility in water, which is 58.2% at 20°C.

In water, the salt hydrolysis yielded is the active form. Stock solutions of potassium sorbate in water can be concentrated up to 50%, which can be mixed with liquid food products or diluted dips and sprays. Sorbates are effective in retarding the growth of many food spoilage organisms. Sorbates have many uses because of their milder taste, greater effectiveness, and broader pH range (up to 6.5), when compared to either benzoate or propianate. Thus, in foods with very low pH, sorbate levels as low as 200 ppm may give more than adequate protection. The solubility of potassium sorbate is 139 g/100 mL at 20°C; it can be applied in beverages, syrups, fruit juices, wines, jellies, jams, salads, pickles, etc.

Sodium benzoate: The use of sodium benzoate as a food preservative has been limited to products that are acid in nature. Therefore, it is mainly used as an antimycotic agent (most yeasts and moulds are inhibited by 0.05-0.1%). The benzoates and parabenzoates have been used primarily in fruit juices, chocolate syrup, candied fruit peel, pie fillings, pickled vegetables, relishes, horseradish, and cheeses. Sodium benzoate is more effective in food systems where the pH is as low as 4.0 or below.

Other additives: Other naturally antimicrobial compounds found in fruits and vegetables include:

— Vanillin (4-hydroxi-3-methoxybenzaldehyde) is found primarily in vanilla beans and in the fruit of orchids (Vanilla planifola, Vanilla pompona, or Vanilla tahitensis). Vanillin is most active against moulds and non-lactic acid gram-positive bacteria. The effectiveness of vanillin against certain moulds such as A. flavus, A. niger, A. ochraceus, or A. parasiticus has been demonstrated in laboratory media, as well as its effectiveness against yeasts such as Saccharomyces cerevisiae,

Pichia membranaefaciens, Zygosaccharomyces bailii, Z. rouxii and Debaryomyces hansenii.

— Allicin is an antimicrobial present in the juice vapour of garlic. This compound is effective in inhibiting the growth of certain pathogenic bacteria such as B. cereus, C. botulinum, E. coli, Salmonellae, Shigellae, S. aureus, A. flavus, Rhodotorula, and Saccharomyces.

— Cinnamon and eugenol are reported to have an inhibitory effect on the spores of Bacillus anthracis. Also, cinnamon was found to inhibit the growth of the aflatoxin of A. parasiticus. Aqueous clove infusions of 0.1 to 1.0% and 0.06% eugenol were reported to inhibit the growth of germinated spores of B. subtilis in nutrient agar.

— Oregano, thyme, and rosemary have been found to have inhibitory activity against certain bacteria and moulds due to the presence of antimicrobial compounds in their essential oils (e.g., terpenes, carvacol, and thymol).

Thermal Treatment

The main function of heat in food processing is to inactivate pathogenic and spoilage organisms, as well as enzyme inactivation to preserve foods and extend shelf life. Other advantages of heat processing include the destruction of anti-nutritional components of foods (e.g., trypsin inhibitors in legumes), improving the digestibility of proteins, gelatinisation of starches, and the release of niacin. Higher temperatures for shorter periods achieved the same shelf life extension as food treated at lower temperatures and longer periods, and allowed retention of sensory and nutritional properties.

Hot water: Hot water plays an important role in the sanitation of food products before processing. Some food products are treated with hot water to eliminate insects, and to inactivate microorganisms and enzymes. Foods are retained in a water blancher at 70-100°C for a specific time and then removed to a dewatering and cooling system.

Steam: Steam is a more effective means than hot water for blanching foods such as fruits and vegetables. This method is especially suitable for foods with large areas of cut surfaces. It retains more soluble compounds and requires smaller volumes of waste for disposal than those from water blanchers. This is particularly so if air-, rather than water-cooling is used. Furthermore, steam blanchers are easier to clean and sterilise.

Temperatures ranging from 10 to 15°C above the optimum temperature for growth will destroy vegetative cells of bacteria, yeasts, and moulds. Most vegetative cells, as well as viruses, are destroyed when subjected to temperatures of 60 to 80°C for an appropriate time. Somewhat higher temperatures may be needed for thermophilic or thermoduric microorganisms. All vegetative cells are killed in 10 min at 100°C and many spores are destroyed in 30 min at 100°C. Some spores, however, will resist heating at 100°C for several hours.

Freezing

Freezing is a quick and convenient way to preserve fruits and vegetables at home. It is a popular method of home food preservation. Home frozen fruits and vegetables of high quality and maximum nutritional value can be produced if the directions below are followed. These directions are based on:

— the chemical and physical reactions which take place during the freezing process;

— scientific knowledge of the effect of freezing on the tissues of fruits and vegetables; and

— food microbiology.

Fresh fruits and vegetables, when harvested, continue to undergo chemical changes which can cause spoilage and deterioration of the product. This is why these products should be frozen as soon after harvest as possible and at their peak degree of ripeness. Fresh produce contains chemical compounds called enzymes which cause the loss of colour, loss of nutrients, flavour changes, and colour changes in frozen fruits and vegetables. These enzymes must be inactivated to prevent such reactions from taking place.

Enzymes in vegetables are inactivated by the blanching process. Blanching is the exposure of the vegetables to boiling water or steam for a brief period of time. The vegetable must then be rapidly cooled in ice water to prevent it from cooking. Contrary to statements in some publications on home freezing, in most cases blanching is absolutely essential for producing quality frozen vegetables. Blanching also helps to destroy microorganisms on the surface of the vegetable and to make some vegetables, such as broccoli and spinach, more compact.

The major problem associated with enzymes in fruits is the development of brown colours and loss of vitamin C. Because fruits are usually served raw, they are not blanched like vegetables. Instead, enzymes in frozen fruit are controlled by using chemical compounds which interfere with deteriorative chemical reactions. The most common control chemical is ascorbic acid (vitamin C). Ascorbic acid may be used in its pure form or in commercial mixtures with sugars.

Some directions for freezing fruits also include temporary measures to control enzyme-activated browning. Such temporary measures include soaking the fruit in dilute vinegar solutions or coating the fruit with sugar and lemon juice. However, these latter methods do not prevent browning as effectively as treatment with ascorbic acid.

Another group of chemical changes that can take place in frozen products is the development of rancid oxidative flavours through contact of the frozen product with air. This problem can be controlled by using a wrapping material which does not permit air to pass into the product. It is also advisable to remove as much air as possible from the freezer bag or container to reduce the amount of air in contact with the product.

Textural Changes

Water makes up over 90 percent of the weight of most fruits and vegetables. This water and other chemical substances are held within the fairly rigid cell walls which give support structure, and texture to the fruit or vegetable. Freezing fruits and vegetables actually consists of freezing the water contained in the plant cells. When the water freezes, it expands and the ice crystals cause the cell walls to rupture.

Consequently, the texture of the produce, when thawed, will be much softer than it was when raw. This textural difference is especially noticeable in products which are usually consumed raw. For example, when a frozen tomato is thawed, it becomes mushy and watery. This explains why celery, lettuce, and tomatoes are not usually frozen and is the reason for the suggestion that frozen fruits, usually consumed raw, be served before they have completely thawed. In the partially thawed state, the effect of freezing on the fruit tissue is less noticeable. Textural changes due to freezing are not as apparent in products which are cooked before eating because cooking also softens cell walls. These changes are also less noticeable in high starch vegetables, such as peas, corn, and lima beans.

The extent of cell wall rupture can be controlled by freezing produce as quickly as possible. In rapid freezing, a large number of small ice crystals are formed. These small ice crystals produce less cell wall rupture than slow freezing which produces only a few large ice crystals. This is why some home freezer manuals recommend that the temperature of the freezer be set at the coldest setting several hours before foods will be placed in the freezer.

Some freezer manuals tell the location of the coldest shelves in the freezer and suggest placing unfrozen products on these shelves. All freezer manuals give guidelines for the maximum number of cubic feet of unfrozen product which can be frozen at one time. This is usually 2 to 3 pounds of vegetable to each cubic foot of freezer space per 24 hours. Overloading the freezer with unfrozen products will result in a long, slow freeze and a poor quality product.

Temperature Fluctuation

To maintain top quality, frozen fruits and vegetables should be stored at 0° F or lower. This temperature is attainable in separate freezer units and in some combination refrigerator-freezers. A freezer thermometer can help you determine the actual temperature of your freezer. Storing frozen foods at temperatures higher than 0° F increases the rate at which deteriorative reactions can take place and can shorten the shelf life of frozen foods.

Do not attempt to save energy in your home by raising the temperature of frozen food storage above 0° F. Fluctuating temperatures in the freezer can cause the migration of water vapour from the product to the surface of the container. This defect is sometimes found in commercially frozen foods which have been improperly handled.

Loss of Moisture

Moisture loss, or ice crystals evaporating from the surface area of a product, produces freezer burn—a grainy, brownish spot where the tissues become dry and tough. This surface freeze-dried area is very likely to develop off flavours. Packaging in heavyweight, moistureproof wrap will prevent freezer burn.

Growth of Microbes

The freezing process does not actually destroy the microorganisms which may be present on fruits and vegetables. While blanching destroys some microorganisms and there is a gradual decline in the number of these microorganisms during freezer storage, sufficient populations are still present to multiply in numbers and cause spoilage of the product when it thaws. For this reason it is necessary to carefully inspect any frozen products which have accidentally thawed by the freezer going off or the freezer door being left open.

Nutrient Value

Freezing, when properly done, is the method of food preservation which may potentially preserve the greatest quantity of nutrients. To maintain top nutritional quality in frozen fruits and vegetables, it is essential to follow directions contained in this leaflet for pretreatment of the vegetables, to store the frozen product at 0°F and to use it within suggested storage times.

Storage Times

Most frozen fruits maintain high quality for 8 to 12 months. Unsweetened fruits lose quality faster than those packed in sugar or sugar syrups.

Most vegetables will maintain high quality for 12 to 18 months at 0° F or lower. However, it is a good idea to plan to use your home frozen vegetables before the next year crop is ready for freezing.

Longer storage of fruits and vegetables than those recommended above will not make the food unfit for use, but will decrease its quality.

Quality of Freezer Containers

You must use good quality freezer containers to maintain the quality of frozen fruits and vegetables. A high quality wrap should be both moisture and vapour proof so that moisture can be kept in the product and air kept away from it. Many moisture- and vapour-resistant wraps, such as heavyweight aluminum foil, plastic coated freezer paper, saran, and other plastic films, are effective at excluding oxygen. They should be strong, pliable, and adhere to the shape of the food item. These can be sealed easily with heat or freezer tape.

Be sure to use only tape that is designated for the freezer because other household tapes lose adhesive quality in the extremely cold freezer temperatures. These wraps are not as convenient for fruits and vegetables as plastic bags or rigid freezer containers. Plastic film bags made especially for freezing are readily available. They seal with twist and tie tops.

Collapsible cardboard freezer boxes are frequently used as an outer covering for plastic bags to protect them against tearing, and for easy stacking in the freezer. Plastic sandwich bags and bread wrappers are not suitable for freezing.

"Freeze-and-cook" bags withstand temperatures from below 0° F to above the boiling point and are suitable for both freezing and cooking the product. These come in 1½ pint and quart sizes and also as large rolls of plastic so that they can be made the size desired. A heat sealer is necessary for closing these bags. These products are more expensive but allow greater convenience.

Packing Methods

There are three ways to pack fruits for freezing: sugar pack, syrup pack, and unsweetened pack. Although some fruits may be packed without sweeteners, the flavour of many fruits is retained better with the use of sugar. Gooseberries, currants, cranberries, blueberries, and rhubarb give good quality packs without or with sugar.

To freeze fruits using sugar pack, sprinkle the required amount of sugar over the fruit. Gently stir until the pieces are coated with sugar and juice. To make sugar syrup, dissolve the needed amount of sugar in cold water. Stir the mixture and let stand until the solution is clear.

There are two basic methods for packing vegetables for freezing, the tray pack and the dry pack.

Dry pack: This is the method used to describe the packing of blanched and drained vegetables into containers or freezer bags. Pack the vegetables tightly to cut down on the amount of air in the container. If the vegetables are packed in freezer bags, press air out of the unfilled part of the bag. When packing broccoli, alternate the heads and stems.

Tray Pack: This is the method of freezing individual pieces of blanched and drained vegetables on a tray or shallow pan, then packing the frozen pieces into a freezer bag or container. This method produces a product similar to commercially frozen plastic bags of individual vegetable pieces and is particularly good for peas, corn, and beans.

In this method it is most important to pack the individually frozen pieces into a bag or container as soon as they are frozen.

Necessary equipment for processing vegetables are:

— a large kettle (minimum capacity of 2 gallons)

— a colander, wire basket, or net bag for blanching

— large pans for cooling

— ice cubes or ice blocks for cooling

— knives

— plastic freezer bags or other containers

— a timer or a clock with a second hand

— hot pads

Choose vegetables for freezing that are at their peak of flavour and texture. If possible, harvest the vegetables in the cool part of the morning and process as quickly as possible. If the freezing process is delayed, immerse the vegetables in very cold water or refrigerate in shallow trays to preserve quality and nutrients. Carefully follow the blanching instructions in the included table for each vegetable. Count the blanching time from when the vegetable is immersed in the vigorously boiling water. The quality of water used to blanch the vegetables can have an effect on the texture of certain vegetables. Very hard water can cause the toughening of vegetables such as green beans. If you have problems with excessively tough green beans, check into the level of hardness in your water supply.

Boiling Water Blanching

— Use 1 gallon water for each pound of vegetable except for leafy greens, which need 2 gallons per pound.

— Bring water to rolling boil.

— Immerse wire basket or blanching basket mesh bag containing vegetable.

— Cover kettle and boil at top heat the required length of time. Begin counting time as soon as you place the vegetable in water. You may use the same blanching water 2 or 3 times. Keep it at required level.

— Cool immediately in ice water for same time used for blanching.

— Drain the vegetables thoroughly. Extra water will form too many ice crystals.

— Pack using dry or tray pack method.

— Freeze.

Steam Blanching

— Put 1 inch of water in kettle, bring to a rolling boil.

— Suspend a thin layer of vegetable in wire basket or loose cheesecloth over rapidly boiling water.

— Cover and steam blanch vegetable required amount of time as listed on table.

— Complete as for boiling water blanching.

Oven Blanching

Some directions are available for microwave blanching of vegetables. Incomplete inactivation of enzymes may occur due to the variability of microwave heating resulting in a shorter storage time in the freezer. Lack of standardisation of power levels among various microwave ovens make it impossible to publish a blanching timetable that can be used with all microwave ovens. Follow the instructions in the manufacturer's microwave cookbook. Vegetables blanched in the microwave should be chilled in ice water and processed as regular frozen vegetables.

To Use Home Frozen Produce

Fruits: Thaw fruit at room temperature in its original package to preserve quality and nutritive value. If faster defrosting is required, submerge (if watertight) in cool or lukewarm water or follow microwave defrosting instructions. Serve as soon as defrosted, preferably while a few ice crystals remain.

Vegetables: All vegetables may be cooked from the frozen state except corn-on-the-cob, which should be partially defrosted. Cook frozen vegetables in a small amount of salted water (about ½ cup or less). Cook only until tender—about half as long as if the same vegetable were fresh. You can use a pressure saucepan for cooking frozen vegetables. Follow manufacturer's directions for cooking time. A pack should be thawed enough to break it up before pressure cooking.

Wax Coating

In recent years, much attention has been paid to explore the potential of surface coating to maintain the quality of harvested fresh produce and to reduce the volume of disposable non-biodegradable packaging materials. Many fruits develop a waxy coat on their epidermis as they mature on the plant (apples, bananas, mangoes, and grapes) they develop a waxy coating on the epidermis, which is either felt or seen as a powdery bloom. Wax usually develops when the fruit has attained 2/3 of its growth. But much extraneous matter and spray residues.

The natural waxy coat is not adequate to offer protection against water loss and high respiration rate. Thus the approach towards prolonging the life of fruit / vegetable involves restricting the rate of respiration and preventing moisture loss. So as to maintain the vital food elements in as near the same quality as in the freshly picked fruit or vegetable.

The growing importance of fruits and vegetables in commerce has led to development in various protective wax coatings with and without fungicides, bactericides, growth regulators, etc., wax coating of fresh fruit and vegetable is quite old and it has great potential in the storage and transportation of fresh produce. Surface coatings have been used extensively on bulky organs to modify internal atmosphere composition and thereby delay ripening, reduce the water loss, and it improves the finish of the skin.

Post harvest losses can be reduced to some extent by, selection of varieties, careful handling, cleaning and sorting, control of ripening process, control of spoilage, packaging, use edible films, use of fungicide, use of ethylene absorbent, use of sprout inhibitors, use of growth regulators, irradiation, packaging home operations, dumping, washing and waxing.

Waxing of fruits and vegetables is done according to the recommendations. Food grade waxes are used to replace some of the natural; waxes removed in washing and cleaning operations, and this helps to reduce the water loss during handling and marketing. If produce is waxed, the wax coating is allowed to dry thoroughly before further handling. Waxing of the produce is an old age art, which was started in the beginning of the 19th century. Waxing was used as a preservation technique for fruits since 1900. Wax has been used since pre historic times.

In 4200 BC, the ancient Egyptians kept bees and used the wax from honeycombs to make models and to preserve their dead and in encaustic painting. The Romans used in pottery making textile printing, as early as 12th century, the Chinese were using waxing processes on fruits, but it was not until 1922 that a waxing process was introduces for wide spread use on fruits and vegetables.

Wax is an ester a long chain aliphatic acid chain aliphatic alcohol. Waxes are the esters formed from a fatty acid and a high molecular weight alcohol. Principle involved during storage: The produce continues respiring and uses up all the O_2 with in the fruit which is not being replaced as quickly as before waxing because of the coating and produces CO_2, which accumulates within the fruit because it cannot escape as easily through the coating. Eventually the fruit will switch to anaerobic respiration that does not require O_2. An aerobic respiration produces much more CO_2, acetaldehyde and ethanol. The acetaldehyde and ethanol give off - flavour to the product, which are detrimental to the perceived quality. The natural barrier of the fruit and vegetable, the type of wax and amount if was applied will influence the extent to which the internal atmosphere (O_2 and CO_2) are modified and the level of reduction in weight loss.

All fruits and vegetable are living organisms and as such require intake of O_2 and release CO_2 for respiration (aerobic). They also release vapour top the atmosphere through diffusion. Both these processes create a loss in weight of the produce. All fruit and vegetables are covered naturally in a cuticle, which is barrier to gas exchange. The cuticle is very impermeable to O_2, slightly more permeable to CO_2 and several magnitudes more permeable to water vapour.

Unwaxed fruit or vegetables have pores in their surface through which most the diffusion of O_2 and CO_2 offers. Water vapour can move through the pores, cuticle and micro-cracks in the cuticle. During the process of waxing a tightly adhering thin film of the coating substance is applied to the surface f the fruit. The surface coating is often relatively impermeable to O_2 and CO_2 and water.

Human became more health conscious. This gave new awareness of fresh fruits and vegetables, and resulted in rapidly increasing demand. Consumption of fresh produce, to help the consumers to obtain possible product, Fruits and vegetables are living organisms, which contain 80 - 90 % water by weight. If they are left without cuticle, the water quickly begins to evaporate, resulting in poor product shelf life to increase freshness and appearance. The fruit and vegetable growers should concern about consumer safety. There are few active ingredients which when combined with water, wetting and drying agents and form the compounds used a protective coating. The coating may contain one or more of the following. Carnauba wax, shellac wax, paraffin wax, candelilla wax, bees wax.

Types of Waxes

Animal waxes

1. Bees wax.
2. Spermaceti wax
3. Shellac wax
4. Chinese insect wax

Vegetable waxes

1. Carnauba wax
2. Candelilla wax
3. Sugarcane wax
4. Palm wax
5. Esparto wax
6. Japan wax
7. Oricury wax
8. Waxol 0.12

Mineral and synthetic wax

1. Ozocerite
2. Montan wax
3. Synthetic wax

Foodstuffs that are waxed are:

Fruits:

Apple Avocadoes Bell pepper Lemon Grapes

Banana Melons Oranges Lime Passion fruit

Peaches Pine apples

Vegetables:

Cucumber Tomato Sweet potato Melon

Methods of Waxing

Principle methods of waxing of fruits and vegetables are:

1. *Liquid Paraffin wax method*: In this method fruits and vegetables are dipped in hot paraffin. Some times resins are added. The main disadvantage of this method is too much of coating material is used.

2. *Slab wax method*: In this case the wax is pressed against rapidly revolving brushes. But the efficiency is very less.
3. *Spray method*: Spraying of melted wax on the fruit, which is subsequently brushed mechanically until a film of desired thickness is obtained. The wax is dissolved in a suitable solvent. This depends on,
   - — The pressure employed
   - — Volume of wax used
   - — Wax temperature
   - — Distance of fruit from the spray
   - — Number of spray nozzles.
4. *Dipping or cold wax method*: Fruits and vegetables are washed and then without being dried are dipped into a wax emulsion of proper concentration. They are dried before packing. Purified wax is odour less, tasteless and nontoxic and it can be heat- sealed.

Properties of Wax

- — Kneadable at 20° C.
- — Easily emulsifiable.
- — Should not impart undesirable odour.
- — Should be economical.
- — Efficient drying performance.
- — Non-sticky or tacky.
- — Should never interfere with the quality of fresh fruit/vegetable.
- — Melts above 40° C without decomposition.
- — Has relatively low viscosity.
- — Capable of being polished by slight pressure.
- — Translucent to opaque form but not like glass.

The properties of wax depend primarily on molecular structure rather than molecular size and chemical constitution.

Functions of Waxing

- — To reduce the weight loss.
- — To replace the natural wax.
- — To increase the appearance

— To increase the freshness

— To decrease the rate of transpiration

— To reduce the shrinkage losses.

— To inhibit mold growth.

— To prevent other physical damage and disease.

During harvesting, transport, washing, grading, it can undergo some scratching or abrasion, which removes natural layer or was and also scars the protective layer of skin.

Lipids and Resins

Lipids and resin are added to coating formulations to reduce gas exchange, but mainly to impart hydrophobicity (to reduce water loss) and gloss appearance. Fruits coated with resins will develop whitening of the skin due to condensation that develops when they are brought from cold storage to ambient conditions.

Composite and Bilayer

Several composite and bilayer films have been investigated with goal of combining the desirable properties of different material to improve permeability characteristics, gloss, strength, flexibility, nutritional value, and general performance or coating formulations. Plasticizers have been incorporated into edible coatings as a processing aid to facilitate coating application and to increase merchantability. The most commonly used plasticizers are polyols (sorbitol and glycerol, mono, di or oligo saccharides, lipids).

The plasticizing effect is due to their ability to reduce internal hydrogen bonding while increase intermolecular spacing. This will reduce the brittleness of the film but increases the permeability of water. Several wax formulations have been developed and tasted experimentally. Suitable emulsifiers are used to produce a wax emulsion. Synthetic or natural resins are added to the wax emulsions to give more gloss to the treated produce. Waterwax emulsions are safer to use than solvent waxes, which are highly flammable.

Water -wax emulsion can be used without drying the fruit prior to the treatment, therefore washing and wax treatment can be combined. Carnauba wax, sugarcane wax, thermoplastic terpene resins and shellac are common substances, which are used commercially; however waxes of plant origin are preferred. Development in wax emulsions were not limited to the formulation of new types of wax, but also included the addition of chemical fungicide, which gave waxing treatment an additional function. For example, biphenyl amine added to the wax emulsion not only gave additional shine and reduced weight loss and respiration rate of apples and pears, but also reduced the occurrence of scald.

Addition of post harvest fungicides to the wax, emulsion to control Penicillium rot in stored citrus fruits also proved helpful. Waxes may be applied by dipping, spraying, foaming or brushing. Foaming is a better method of application since it leaves a very thin layer of wax on the fruit. If a thicker layer of wax is applied to the fruit surface, it becomes an undesirable barrier between the external and internal atmosphere and restricts exchange of respiratory gases (CO_2 and O_2). This may result in anaerobic respiration, resulting in fermentation and development of an off flavour. It is therefore necessary to adjust the thickness of the wax coat according to the variety and storage and marketing temperatures.

Effect of Wax Coatings

Wax Coating of fruit improved its external appearance by giving additional shine to the fruit surface. It reduced weight loss and kept the fruit firmer, thus marinating its fresh look, while it reduced rates of respiration and ethylene production, thus delaying senescence. No significant change in constituents like Vitamin C, acidity, and sugar contents (In citrus fruit especially) were recorded due to treatment. However higher concentration of wax resulted in the development of off flavour in citrus fruits particularly during holding at 20°C, making this treatment unsuitable for this fruit. It was, however found necessary to tailor the thickness of the wax coat according to storage temperature to avoid development of off flavour.

FDA regulates these waxes, or coatings, as food additives approved or "generally recognised as safe" for human consumption. However, some consumers have concerns about their use. Vegetarians and others who avoid animal products may worry that fruits and vegetables contain animal-based waxes, such as oleic acid. Some people fear that the wax traps pesticides, making the fruit or vegetable unsafe to eat-even though FDA's pesticide monitoring programme indicates that pesticide residues on fruits and vegetables are consistently within acceptable safe limits.

If consumers want to avoid waxed fruits and vegetables, FDA regulations that took effect in 1994 may help them identify the appropriate products for them. These regulations require produce packers or grocers to provide point-of-sale information about the presence of waxes on fresh fruits and vegetables. This information can appear on labels of individual products, packing cartons (if they are used at the point of sale), or on counter cards or signs. The information will say that the product is: Coated with food-grade animal-based wax to maintain freshness, or Coated with food-grade vegetable-, petroleum-, beeswax-, and/or shellacbased wax or resin to maintain freshness.

If only one of these types of waxes is applied, the label can simply identify the type, such as "vegetable-based." FDA also will allow the statement "No wax or resin coating" on fresh fruits and vegetables that do not contain wax. Besides reading labelling information, consumers can reduce their concerns about waxes by rinsing fruits and vegetables with warm water and, when appropriate, scrubbing with a brush. This will eliminate much of the wax.

Drying Fruits and Vegetables

Drying is the oldest known method of preserving food. Historically, food was dried in the sun. Nowadays, we can sun dry or dry in an oven or a dehydrator that is especially designed for home drying. Drying has many advantages. It is safe. By removing most of the moisture from a food, microorganisms can't grow in it. Dried food is compact so that less storage space is needed. Plus, dried foods keep well, travel well and add variety to our diets.

Fruits and vegetables selected for drying should be of the highest quality—fresh and ripe. Drying does not improve the quality. Immature produce will lack flavour and colour. Overmature produce may be fibrous or mushy. If the food is not perfect for eating, it is not suitable for drying.

Dried vegetables can be used as chips, reconstituted for a cooked side dish, used in soups, stews, casseroles and stuffings or made into powders. Dried fruits tend to be chewy and make delicious snacks. Pieces of dried fruits are good in cookies, muffins, cakes and reads. They can also be reconstituted and used in sauces, pies or cobblers or added to gelatin salads, cooked cereals and ice cream.

Foods can be dried in an electric dehydrator, in the sun, in a solar dryer, or in a regular oven.

Dehydrator Drying

Dehydrator drying produces the best quality dried products, so it's not surprising that it's also the most popular drying method. Dehydrator drying also gives you greater flexibility than other methods because it does not depend on dry, sunny days or take over your oven. A variety of electric dehydrators are available for purchase. A dehydrator should have a heat source, a thermostat, and some method of air circulation. If you buy a dehydrator, follow the directions that come with it. Before you buy a new or used food dehydrator, check to see that it has all these features:

— Instruction manual.

— Thermostatically controlled temperature dial with settings between 130° and 150°F. (If you plan to dry meat jerky in your dehydrator, the dehydrator must be capable of maintaining a temperature of 145°. Contact the extension educator in your county for instructions on how to safely prepare meat jerky.)

— Fan or blower to distribute warm air evenly. Shelves made of stainless steel or food-grade plastic. (Galvanized screening is not food-safe.)

— Easy loading and unloading features.

— Outside cabinets made of hard plastic, aluminum, or steel. The highest quality dehydrator has double-wall construction with insulating material sandwiched between the walls to reduce the amount of heat lost during use.

— Enclosed heating element. Appropriate number of trays for your use. Most food dryers come with 4 to 10 food trays.

— Source of replacement parts.

Sun Drying

Sun drying works best when the temperature is in the 90s, the humidity is low, and air pollution levels are low. A major advantage to sun drying is its low cost. Drying trays, netting to protect against bugs, and food to dry are your only investments. Sun drying makes you dependent on the weather, however. If it is sunny one day and cloudy the next, you will have to finish drying by another method. That's because spoilage can occur while the drying food still has enough moisture for microbial growth. Also, you should bring the food inside on cool nights. Another disadvantage is time. What dries in 6 to 8 hours in an electric dehydrator may take 2 to 4 days in the sun.

Solar Drying

Solar drying is like sun drying, only better. The sun's rays collect in a solar box so that, compared with sun drying, drying temperature is higher and drying time is shorter. The shorter drying time gives microorganisms less chance to cause spoilage. If you do not want to buy or build a solar box, you can use the back window ledge of an automobile where the sun shines through. Crack the windows slightly to allow air flow so temperatures do not get too hot. Cover the trays with netting to keep bugs out. Oven Drying You can use your oven to dry small amounts of food at one time. You'll have little or no investment in equipment and you won't have to depend on the weather. Although oven drying produces a safe, generally tasty product, don't expect top quality. Oven-dried food is more brittle and usually darker and less flavourful than food dried in a dehydrator. Another disadvantage of oven drying is its energy cost.

Oven drying takes two or three times longer than drying in a dehydrator. Before drying in an oven, test the oven temperature with an oven thermometer for about 1 hour. Prop open the oven door as you would when actually drying fruit. The oven should maintain a temperature of 130° to 150°F.If the oven cannot maintain a temperature in this range, you will not have high-quality dried food. If the oven is too hot, your food will begin to cook instead of dry. If it is too cool, your food may not dry fast enough and spoil instead.

If you're new to drying, start with the fruits you like best. Think also about how you will use your dried fruits. Peaches or pears in a tangy stewed fruit? Apples or apricots for lunch box snacks? Berries to toss into muffin or cake batters? Most fruits are easy to dry High-quality fruits make the best dried products. Choose firm, fully ripe fruit that is heavy for its size. Handle fruits gently and process them immediately because fruit ready for drying is very fragile. Use overripe or bruised fruits in other ways (for example, as fruit leathers).

Vegetables for drying should be fresh, tender, and just mature. Avoid immature vegetables because their colour and flavour tend to be weak or poor. Also avoid excessively mature vegetables, which are inclined to be tough, woody, or fibrous. For the best quality and nutrition, dry vegetables as soon as possible after harvest.

Preparation of Dried Fruits

Fruits

Gently wash all fruits in cold water just before drying to remove dirt, bacteria, and insects. Thoroughly wash fruits that have skins you will not peel off, such as cherries and prunes. Do not soak fruit because extended soaking can cause nutrient loss and waterlog the fruit, which increases drying times. Remove fruit stems and peels. Peels may be left on some fruits, such as apples and peaches, but they may become bitter or discolour during drying. Core or pit the fruit and cut it into uniform halves, quarters, or slices. Trim away diseased or soft spots.

Vegetables

Wash vegetables in cold water just before drying. If vegetables are covered with soil, wash them under clean running water to prevent the dirt from resettling on the food. Do not allow vegetables to soak in water. Most vegetables should be peeled and trimmed then cut, sliced, or shredded into uniform pieces. Although peeling some vegetables such as young zucchini and well-washed carrots is optional, unpeeled vegetables tend to be tougher when dried. Remove fibrous or woody portions and damaged areas.

You can prepare pieces with a food slicer or food processor. Although you can dry and store many foods without pretreatment, pretreatment generally improves quality, particularly for vegetables. Five major reasons for treating foods before drying are to:

— Preserve colour and flavour

— Minimise nutrient loss

— Stop decomposition (enzyme action)

— Ensure more even drying

— Extend storage life

Pretreating Fruits and Vegetables

Decomposition from enzyme action during storage is less a problem with fruits than it is with vegetables. Fruits have higher levels of sugar and acid, which counteract enzyme action. Although pretreating fruit is not necessary, you can use an ascorbic acid/citric acid dip, a salt solution dip, syrup blanching, a honey dip, or a sulphiting procedure. Certain fruits, such as apricots, pears, peaches, and some varieties of apples, tend to discolour with drying. Pretreating those fruits can decrease browning during processing and storage and lower losses of flavour and of vitamins A and C.

If you use a pretreatment method that requires soaking fruits in a water solution, you will need to increase drying time because the fruit will absorb some water. Do not allow foods to soak more than 1 hour. Ascorbic Acid/ Citric Acid Dips. Ascorbic acid/citric acid dips are often used as a pretreatment for fruits. They prevent fruits such as apples, pears, peaches, and apricots from turning brown when cut and exposed to air. An ascorbic acid dip also increases the vitamin C content of the dried fruit.

Use U.S.P. ascorbic acid or food-grade ascorbic acid, which are seasonally available among canning supplies in supermarkets. Vitamin C tablets can also be used. To prepare an ascorbic acid solution, combine ½ teaspoon of ascorbic acid crystals, or three crushed, 500-milligram tablets of vitamin C, with 1 quart water. Stir until the ascorbic acid dissolves. Place the cut fruit in the ascorbic acid solution. Stir the fruit to ensure even coating. Leave the fruit in the ascorbic acid solution for about 5 minutes. Approximately 1 quart of solution will treat 8 cups of fruit.

Pineapple juice or juice from citrus fruits such as oranges, lemons, or grapefruit can also be used as a pretreatment. These juices contain a mixture of citric and ascorbic acids. However, citric acid is a weaker acid than ascorbic acid and is less effective as a pretreatment. You can also use a commercial pretreatment such as the anti-darkening powders often sold with food preservation supplies. Follow the label directions.

Salt Solution Dip. Prepare a solution of 2 to 4 tablespoons of salt per gallon of water. Soak fruit for 2 to 5 minutes, and then drain it well.

Syrup Blanching. Prepare fruit for drying. Prepare a sugar syrup made with 1 part sugar and 2 parts water. If desired, use less sugar. Bring the syrup solution to a boil. Add the fruit, simmer for 5 minutes, then drain the fruit. Place the fruit on drying trays and dry. This fruit product is like a candied fruit.

Honey Dip. A honey treatment for fruit can effectively minimise browning and softening in light-coloured fruit. Prepare a honey-water dip using 1 part honey to 4 parts water. Dip the fruit in the honey solution immediately after slicing, let it soak for about 5 minutes, and drain well. The dried fruit will have a slight honey taste.

Sulphiting. Sulphur dioxide treatments, either sulphiting or sulphuring, are very effective for retarding oxidation and browning in fruit. Fruit flavour and storage life may also improve. Almost all commercially produced light-coloured fruits, such as dried apples, pears, and apricots, are treated with sulphur compounds. However some people have severe allergic responses to sulphur compounds. They should not eat or work with dried fruit pretreated with sulphur or sulphite compounds.

Sulphuring, a complicated and potentially dangerous procedure, is no longer recommended. Sulphiting involves preparing a solution of water and a sulphiting agent and then soaking the cut fruit in the solution. In the United States six sulphur compounds (sulphur dioxide, sodium sulphite, sodium bisulphite, potassium bisulphite, sodium metabisulphite, and potassium metabisulphite) have been listed by the U.S. Food and Drug Administration (FDA) as "Generally Recognised as Safe" (GRAS).

The most popular sulphiting agents for home drying are sodium bisulphite, sodium sulphite, and sodium metabisulphite. They should be either U.S.P. (food grade) or reagent grade (pure). They are available at most wine-making supply centers and some larger supermarkets.

	Amount of sulphur to add per quart of water
Sodium bisulphite	½ to 1 teaspoon
Sodium sulphite	1 to 2 teaspoons
Sodium metabisulphite	1 to 3 teaspoons

The sulphiting process has two steps:

— Prepare the sulphiting solution in a large glass container just before use. Place the cut fruit in the solution. Do not leave the fruit in the sulphiting solution too long or the fruit will be mushy. Use about 10 minutes for sliced fruit and 30 minutes for halved fruit. Do not exceed the recommended quantities of sulphites or soak times.

— After sulphiting, remove the fruit and drain it well. Some people recommend a quick rinse in cold water before drying. Place sulphited fruit on drying trays and dry. Drying times for sulphited fruits are longer because the fruit absorbs some water during soaking.

Blanching (heating in boiling water or steam) is the pretreatment method of choice for vegetables. Almost all vegetables should be blanched before drying to destroy the enzymes that make vegetables deteriorate. Blanching keeps vegetables from browning, becoming bitter, or developing off flavours. Blanching also cleans and softens vegetables and makes them easier to rehydrate later. Although you can use either boiling water or steam for blanching, vegetables lose more nutrients during boiling.

Steam Blanching. Use a steamer or make a steamer out of a kettle with a tight-fitting lid. Place a colander, wire basket, or sieve inside the kettle. Make sure the food will be above the water level. Add 2 inches of water to the kettle and heat it to boiling. Place the container with the loosely packed food in the steamer, cover the kettle tightly, and continue boiling.

Water Blanching. Fill a kettle with enough water to cover the food. Bring the water to a rolling boil and gradually stir in the food. Cover the kettle tightly and boil. You can reuse the water when blanching more of the same food, adding more water as necessary. If the water appears dirty, replace it with clean water.

Determining Blanching Times. Blanching times vary with altitude (higher altitudes require longer blanching times), the type and texture of the vegetable, the amount of vegetable, and the thickness of the pieces. Generally, vegetables should feel and taste firm yet tender. They should not be fully cooked, but they should be heated all the way through. Test the food by cutting through a piece. If sufficiently blanched, it will appear cooked (translucent) nearly to the centre. Underblanching may cause deterioration in storage, poor rehydration, or bad colour. Overblanching makes vegetables lose colour, flavour, and nutrients and gives them poor texture after rehydration.

After Blanching. Drain vegetables by pouring them directly on the drying trays. If you plan to reuse the water, place a large pan under the trays. Wipe the bottom of the drying tray with a clean towel to remove excess water. Draining the vegetables on one tray and then transferring them to the drying tray results in unnecessary handling. Immediately transfer the blanched vegetables into the dehydrator so drying can begin while the vegetables are still warm.

Different foods can be dried at the same time, but try to choose foods that will dry in about the same amount of time. (Dry similarly sized pieces together.) Onions, peppers, and other strong foods tend to flavour other foods, so dry them separately. Moisture must be removed from the food as quickly as possible at a temperature that does not seriously impair the flavour, texture, or colour of the food.

If the temperature is too low at the beginning, the food may spoil before it dries. If the temperature is too high, the surface may harden so that the interior dries much more slowly. Start the dryer at 140° to 150°F, with the exceptions noted in the drying guidelines. After 2 to 3 three hours, lower the dryer temperature to 130°F to 140°F. Adequate air flow can reduce drying times. Monitor the drying process. If necessary, rotate the trays to ensure even drying. You may need to stir grated, shredded, or finely cut foods.

Drying Time

Many factors affect drying time, including type of food, size and moisture content of the food pieces, pretreatment method, dryer type, dryer temperature, relative humidity of the air, and amount of air movement in the dryer and in the surroundings. With so many factors at work, it's impossible to give precise drying times. Generally, you can figure on drying times of 6 to 36 hours for fruit and 3 to 16 hours for vegetables, which take less time due to their lower sugar contents.

In the end, you need to decide when food is dry. Vegetables are sufficiently dry when they are brittle or leathery. Leathery vegetables will be pliable and spring back if folded. Brittle vegetables such as corn and peas will shatter when hit with a hammer. Fruits are sufficiently dry when they are pliable and leatherlike and have no pockets of moisture. Herbs are sufficiently dry when brittle. Their leaves will shatter when rubbed together.

When you think the food is sufficiently dry, remove a piece and allow it to cool completely. Then check for dryness. When you are in doubt about the dryness of a food, continue to dry it. Foods dry more quickly toward the end of the drying period, so check them frequently, and avoid leaving them in the dryer after they are done. Leaving them in will reduce their quality.

Fruit Leathers

Fruit leathers are a wonderful way to use small quantities of fruit or extra-ripe fruit. Fruit leathers, also known as fruit paper and fruit taffies, are chewy fruit roll-ups made from either cooked or uncooked fruit purée. Fruit leathers allow for individual creativity through combinations of different fruits. Generally you can use any kind of fruit, including apples, bananas, berries, grapes, mangos, papayas, peaches, pears, pineapple, plums, and even tomatoes. Citrus fruits alone are generally not recommended. Fruit leathers are an excellent use for slightly overripe or bruised fruit that would otherwise be discarded.

Uncooked Fruit Leathers

1. Select ripe or overripe fruit or fruit combinations.
2. Wash fruit and cut away blemishes. Remove stones or pits. Remove larger seeds from berries, grapes, and tomatoes if you wish. Peel all tough-skinned fruits; peel others if you wish.
3. Cut fruit into chunks and place them in a food chopper, blender, or food processor.
4. Add 1 tablespoon lemon or other citrus juice per quart of yellow or light-coloured fruit, if desired, for keeping fruit colour.
5. Chop, grind, or blend the fruit into a thick purée. If the fruit has little juice, add several spoonfuls of water or fruit juice to obtain a uniform purée. (If uncooked fruit purée is too juicy, it can be cooked to remove excess liquid.)
6. (Optional) Add sugar, honey, or corn syrup to taste. (Generally no additional sweetener is needed, particularly with ripe fruit, because fruit tastes sweeter after being dried.)
7. (Optional) Add spices (for example, cinnamon, nutmeg, cloves, allspice) to taste. Start with ½ teaspoon dried spice per quart of puréed fruit; spice flavours intensify during drying.
8. Use a drying tray designed for fruit leather or line a portion of a drying tray with lightly oiled heavy plastic wrap. Do not completely cover the tray with plastic wrap or the air will be unable to circulate to other trays. Pour a small amount of purée onto the lining wrap. Make sure the tray has an edge to prevent spillage. Tilt the tray until the purée spreads no more than ¼ inch thick almost to the edge of the plastic wrap. (Two cups of purée will cover a 12- by 17-inch drying tray.)
9. (Optional) Sprinkle the purée with chopped nuts, seeds, or grated coconut.

Cooked Fruit Leathers (Double-Boiler Method)

1. Select, wash, and prepare fruit as described for uncooked fruit leather.
2. Cut the fruit into slices or chunks and place them in the top of a double boiler.
3. Add water to the bottom of the double boiler. Cover the double boiler and steam the fruit for 15 minutes or until fruit is soft. (If a double boiler is not available, you can place a small pan containing the fruit in a larger pan that is partially filled with boiling water.)
4. Follow steps 4-9 for uncooked fruit leather.

Cooked Fruit Leathers (Microwave Method)

1. Select, wash, and prepare fruit as described for uncooked fruit leather.
2. Cut the fruit into slices or chunks and place them in a microwavesafe bowl.
3. Cover the bowl with plastic wrap or a microwave lid.
4. Microwave on high for 4 minutes, stir, and rotate the bowl. Continue stirring and rotating the bowl every 4 minutes until fruit is soft.
5. Follow steps 4-9 for uncooked fruit leather.

Fruit Leather from Canned Fruit

1. Thoroughly drain home-canned or commercially canned fruit or use baby food fruit without tapioca.
2. Follow steps 3 and 5-9 for uncooked fruit leather. Since canned fruits have been heat processed to stop enzymatic action, you don't need to add ascorbic acid.

Drying and Storing

Place prepared trays in the dehydrator. Dry until the leather is sticky, generally 6 to 8 hours at 140°F. Properly dried fruit leather will be translucent and slightly tacky to the touch but will still peel away from the plastic wrap. Lift an edge of the leather, which should stick tightly to the surface, and peel it back slightly. If the leather peels away readily, it is dry. If the leather has cooled, it may need to be warmed slightly for a few minutes to help it peel away.

If the fruit leather cracks or chips, it has dried too long, but it is still edible. Remove the remainder of the plastic wrap. If part of the leather is still sticky, you can dry it more without the plastic wrap. After drying the fruit leather, leave it whole or cut it into pieces. The fruit leather can be rolled and wrapped in plastic wrap or stored flat in sheets with plastic wrap separating each sheet. Place the wrapped pieces in an air-tight container in a cool, dark, dry place. You can also store fruit leather in the refrigerator or freezer.

References

Food and Agriculture Organization of the United Nations. 1989. *Prevention of food losses: fruit, vegetable and root crops: A training manual.* Rome: FAO, 157 pp.

Mitchell, F.G., Guillon, R., R.A.. Parsons. 1972. *Commercial Cooling of Fruits and Vegetables.* U.C. Extension Manual 43, University of California, Division of Agricultural Sciences, Oakland, California 94608.

Rangarajan, A, et al. (no date). *Food Safety Begins on the Farm. A Grower's Guide: Good Agricultural Practices for Fresh Fruits and Vegetables.* Ithaca: Cornell University. 28 pp.

Shewfelt, R.L. 1986. Postharvest treatment for extending shelf life of fruits and vegetables. *Food Technology* 40(5):7078, 89.

11

Quality Assurance

There is a world tendency towards a greater consumption of fruits and vegetables due to a growing concern for a more balanced diet, with a lower proportion of carbohydrates, fats and oils and with a higher proportion of dietary fiber, vitamins, and minerals. Another aspect that deserves attention is the tendency towards simplification in the task of preparing daily meals. In the United States, until the 60s, the preparation of lunch or dinner required about 2 hours and was planned in advance.

Nowadays, meals are prepared in less than one hour and the menu to be served at dinner begins to be defined after 4 p.m. The expanding incorporation of processed fruits and vegetables and other ready-made foods are partly responsible for this reduction in the time dedicated to cooking. Probably, the most significant fact that encourages this tendency is woman's increasing incorporation in full-time work that reduces her time to buy and to prepare foods but giving her more capacity to spend money.

Also influencing the consumption patterns is the increasing market segmentation through the expansion in shapes, colours, flavours, ways of preparation, and/or packaging in which a product is presented. Among others, tomatoes are an example of it, since today they can be purchased in at least 4 different types: conventional or "beef tomato", "extended shelf life", "cherry", and processing types sold fresh, all of them in different sizes, packages and in some cases, colour.

There is also an increasing supply of exotic or non-conventional fruits and vegetables, which together with the previous point, notably expands the purchase options. For example, in 1981, in a well-supplied supermarket of the USA, there were 133 options of different fruits and vegetables, but they increased to 282 in 1993 and to 340 in 1995. Without reaching these levels, the same tendency is observed in the different countries of Latin America and the Caribbean.

Lastly, there is a growing demand for higher quality, external as well as internal quality. External aspects (presentation, appearance, uniformity, ripeness, and freshness) are the main components in the decision to purchase, which is usually taken when the consumer sees the product exhibited at the sales point. This is particularly important in the self-service systems where the product must "self-sell" and if it is not chosen, represents a loss for the retailer. Internal quality (flavour, aroma, texture, nutritional value, and absence of biotic and non-biotic contaminants) is linked to aspects not generally perceived externally, but are equally important to many consumers.

In terms of the service or satisfaction that it produces to consumers, we could also define it as the "degree of fulfilment of a number of conditions that determine its acceptance by the consumer". Here, a subjective aspect is introduced, since different consumers will judge the same product according to their personal preferences.

The destination or use can also determine different criteria for judging quality within the same crop. For example, the tomato for fresh consumption is valued essentially by its uniformity, ripeness, and absence of defects, while colour, viscosity, and industrial yield as raw material define the quality for ketchup tomatoes. It is common to use additional words to define the quality to the specific use, such as "industrial quality", "nutritional quality", "export quality", "edible quality", etc.

Quality: A Complex Perception

Quality is a complex perception of many attributes that are simultaneously evaluated by the consumer either objective or subjectively. The brain processes the information received by sight, smell, and touch and instantly compares or associates it with past experiences or with textures, aromas, and flavours stored in its memory. For example, just by looking at the colour, the consumer knows that a fruit is unripe and that it does not have good flavour, texture or aroma. If colour is not enough to evaluate ripeness, he/she uses the hands to judge firmness or other perceptible characteristics. The aroma is a less used parameter except in those cases where it is directly associated to ripeness like in melon or pineapple. This comparative process does not take place when the consumer faces, for the first time, an exotic fruit whose characteristics are unknown.

Fruits and vegetables are consumed mainly for their nutritive value as well as by the variety of shapes, colours, and flavours that make them attractive for food preparation. When they are consumed raw or with very little preparation, the consumer's main concern is that they must be free of biotic or non-biotic contaminants that may affect health.

Elements of Quality

Appearance

Appearance is the first impression that the consumer receives and the most important component of the acceptance and eventually of the purchase decision. Different studies indicate that almost 40% of the consumers decide what to buy inside the supermarket. Shape is one of the subcomponents more easily perceived, although in general, it is not a decisive aspect of quality, except in case of deformations or morphological defects. In some cases, shape is a ripeness index and therefore an indication of flavour. This is the case of the "full cheek" in mango or the "finger" angularity in bananas.

In those species where the inflorescence is the marketable organ such as broccoli or cauliflower or those that form "heads" like lettuce, cabbage, endive, etc. the compactness is the most relevant feature. In general, it is not associated to their organoleptic characteristics but rather is an indicator of the degree of development at harvest, since open inflorescences indicate that they were picked too late while non-compact "heads" are the consequence of a premature harvest. To a certain extent, it is also an indicator of freshness since compactness decreases with dehydration.

Uniformity is a concept applied to all the components of quality (size, form, colour, ripeness, compactness, etc.). For the consumer it is a relevant feature that indicates that someone that knows the product has already selected and separated it into categories based on the official standards of quality. It is so important that making products uniform is the main activity in preparation for the market.

In many cases, internal or external defects do not affect product excellence, but the consumer rejects them since the absence of defects is one of the main components of appearance and therefore, of the primary decision to purchase. Different causes during growth (climate, irrigation, soil, variety, fertilisation, etc.) can lead to morphological or physiological defects. Some examples of the first ones are "doubles" in cherry, root ramifications in carrots, "catface" tomatoes, "knobby" tubers and "hollow heart" in potatoes, etc.

Tipburn on leafy vegetables and black heart in celery due to calcium deficiencies as well as the internal rots in various species due to boron deficiencies are examples of physiological defects. More serious are those physical or physiological defects that originate during or after preparation for the fresh market and that show up at retail or consumer's level. Within the first ones we have the mechanical damages, bruises or wounds that take place during the handling of the product and that are the entrance doors to most pathogens causing postharvest rots. Chilling injury, ethylene effects as well as sprouting and rooting, are physiological responses to inadequate storage conditions.

Freshness and ripeness are part of the appearance and they have components of their own. They are also indicative of the expected flavour and aroma when products are consumed. "Freshness" is the condition of being fresh or as close to harvest as possible. It is used in vegetables where harvest is the point of maximum organoleptic quality characterised by the greatest turgidity, colour, flavour, and crispness.

"Ripeness" is a concept used in fruits that also refers to the point of maximum eatable quality but that in many cases is reached at the level of sales point or of consumption since, in most commercial operations, fruits are harvested slightly immature. For example, fruits stored in controlled atmospheres reach their eatable quality after leaving the store room, several months after harvest.

Within the parameters for defining freshness and ripeness, colour, both intensity and uniformity, is the external aspect more easily evaluated by the consumer. It is decisive in those products like leafy vegetables or unripe fruits such as cucumber, snapbeans, and others where an intense green is associated with freshness and pale green one or yellowing to senescence. Colour is also an indicator of fruit ripeness and very important in those where no substantial changes take place after harvest (non-climacteric fruits), such as citrus, pepper, eggplant, and cucurbits in general. In fruits that suffer changes after harvest (climacteric) colour is less decisive and basically indicates the degree of ripeness, as for example tomato, pear, banana, etc.

Consumers assign to size a certain importance and at equivalent quality, intermediate sizes are preferred. In fruits that are naturally large such as pumpkins, watermelons, melons, etc., there is a very defined trend towards sizes that can be consumed by a family (1-2 kg) in a relatively short period (1 week). Size is one of the main indicators of the moment of harvest and in many cases it is directly associated to other aspects of quality such as flavour or texture. Such is the case of zucchinis, peas, haricot beans and miniature vegetables in general where consumers particularly value small sizes.

Gloss enhances the colour of most products, but it is particularly valued in species like apple, pepper, eggplant, tomato, grapes, plums, cherries, etc., to such a point that many of them are waxed and polished to improve their shine. In vegetables, gloss is associated in a certain way to turgidity: a brilliant green is one of the indicators of freshness. It can also be used as a harvest index in eggplants, cucumbers, squash and other fruits that are harvested unripe where the decrease in shine indicates that they have developed too much and have lost part of their characteristics of flavour and texture. On the contrary, in melon, avocado, and other species, it is indicative that it has reached ripeness for harvest.

Different sensations perceived by humans are included within the definition of texture. Thus, firmness is perceived with the hands and, together with the lips, the kind of food surface (hairy, waxy, smooth, rough, etc.), while teeth determine the rigidity of the structure that has been chewed. The tongue and the rest of the mouth cavity detect the type of particles that are crushed by teeth (soft, creamy, dry, juicy, etc.). Also the ears contribute to the sensation of texture, for example, the noises generated when chewing in those species where crispness is an important aspect.

Together with flavour and aroma, texture constitutes the eating quality. An over-ripe tomato, for example, is mainly rejected by its softening and not because important changes in the flavour or aroma have taken place. Although it is decisive for the quality of some fruits and vegetables, in others it has a relative importance. In texture terms, each product is valued differently: either for its firmness (tomato, pepper), the absence of fibers (asparagus, globe artichoke), its softness (banana), juiciness (plums, pears, citrus), crispness (celery, carrot, apple), etc.

Firmness and colour are the main parameters to estimate the degree of ripeness of a fruit since this process initially improves and softens fruit texture, which together with the changes in flavour and colour, bring the fruit to reach its maximum eatable quality. However, as this process continues, over-ripeness takes place, which leads finally to tissue disorganisation and decay of the product. Firmness is used mainly as a harvest index and it is measured with instruments that register the force necessary for a certain deformation or resistance to the penetration of a piston of known dimensions.

Juiciness is the sensation of liquid spilling inside the mouth as tissues are chewed. The juice content of many fruits increases as they ripen in the plant. It is regulated that the minimum content that citrus fruits should have, is: 30% for Navel oranges; 35% for grapefruit and the other oranges; 25% for lemons; 33% for mandarins and 40% for clementines.

Flavour

Flavour is the combination of the sensations perceived by the tongue (taste) and by the nose (aromas). Although those sensations can be perfectly separated one from the other, as the sensitive receptors are so close, simultaneously with the act of bringing near the mouth, of biting, chewing, and tasting, we are perceiving the aromas, particularly those that are liberated with the crushing of tissues. It is also possible, however, that certain external aspects (particularly ripeness) anticipate the flavour that should be expected when consuming the product. The human being has stored in his/her memory an enormous quantity of different tastes and aromas and, if it was eaten previously, is able to recognise them without seeing the product.

In fruits and vegetables, taste is usually expressed in terms of the combination of sweet and sour principles that are an indication of ripeness and eating quality. The content of soluble solids is a good estimate of total sugar content, and many fruits should have a minimum content of solids to be harvested. Organic acids (citric, malic, oxalic, tartaric) are the other important components of taste, particularly in their relationship with soluble solids.

As the fruit ripens they tend to diminish and so the relationship with the soluble solids tends to increase. Titratable acidity is the form of expressing acidity. The soluble solids/titratable acidity relationship is a denominated ratio and it is essentially used in citrus where it is a function of the species and of the variety. Its value is 8 for mandarins, Navel oranges, and hybrids, 7 for other type of oranges, and 5.5 for grapefruits.

Astringency and bitter tastes are due to different compounds. They are not frequent and when they exist, usually diminish with ripening. In those cases in which they appear naturally and represent a disadvantage, they have been eliminated through breeding programmes.

There are specific compounds that characterise certain species or a group of them, for example, pungency in the peppers denominated "hot" is basically determined by the capsaicin content and other 4 structurally similar compounds. There are also cases in which enzymes and substrata responsible for the taste are compartmentalised in healthy tissues and they only get in contact by cutting, chewing or crushing.

This is the case of pungency in garlic and onion and also of the taste of raw cucumber. Cooking these vegetables whole prevents these reactions and the resulting taste is different.

Table 1. Recommended minimum soluble solid content at harvest.

Apple	10,5-12,5
Apricot	10
Blueberries	10
Cherry	14-16
Grape	14-17,5
Grapefruit	8
Kiwifruit	14
Mango	8
Mandarin	8
Melon	10-12
Nectarine	10
Orange	8
Papaya	11,5
Peach	10
Pear	13
Persimmon	18
Pineapple	12
Plum	12
Pomegranate	17
Raspberry	8
Strawberry	7
Watermelon	10

There is a correlation between dry matter content and organoleptic characteristics mainly used by the industry. In general, a higher content of solids means higher industrial yield and taste. This is particularly important in dehydrated products. In potato, a higher content of dry matter (measured as specific gravity) is associated to a better cooking quality. For the fresh market, however, dry matter content is not used as an indicator of the time of harvest and/or organoleptic quality, except in the case of avocado where there is a close correlation with the oil content. Depending on the variety considered, avocados with a dry matter content lower than 21-23% should not be marketed.

The aroma of fruits and vegetables is due to the human perception of numerous volatile substances. Refrigerated fruits and vegetables are less aromatic since volatile liberation diminishes with temperature. As well as in the case of taste, many aromas are liberated when tissues lose their integrity.

Nutritive value

From the nutrition point of view, fruits and vegetables are insufficient to satisfy daily nutritional requirements, essentially because of their low content of dry matter. They have a high content of water and are low in carbohydrates (excepting sweet potatoes, potatoes, cassava, and other underground organs), proteins (except for legumes and some crucifers) and lipids (except avocados), but they are, in general, a good source of minerals and vitamins. Different countries have made tables of recommended daily consumption

The discovery that certain foods have biologically active compounds, beneficial to health beyond basic nutrition opened a new stage in nutrition science. These compounds or their metabolites that have been denominated "functional", help to prevent diseases like cancer, have a protective effect on cardiovascular problems, are neutralisers of free radicals, reduce cholesterol and hypertension, prevent thrombosis, besides other beneficial effects.

Those foods that contain them are also called "functional" although other names such as "pharmafoods", "nutraceuticals", and others have been proposed. As most of these compounds are of plant origin, many authors call them phytochemicals. Fruits and vegetables are rich in phytochemicals like terpenes (carotenoids in yellow, orange and red fruits and limonoids in citrus), phenols (blue, red and purple colours of cherries, grapes, eggplant, berries, apples and plums), lignans (broccoli), thiols (sulfur compounds present in garlic, onion, leek and other alliums, cabbages and other crucifers) and others.

Safety

Fruits and vegetables should be attractive, fresh, nutritive and with a good appearance and presentation. Besides these characteristics, their consumption should not put health at risk. The consumer has no way to detect the presence of dangerous substances on food and he/she depends entirely on the seriousness and responsibility of all the members of the production and distribution chain. Necessarily, he/she has to trust them, in addition to the usual precautions of washing, peeling and/or cooking the product before consuming it.

However, this trust is very volatile and any suspicion about safety has a tremendous impact at consumer level. Among the most relevant examples it is worth to mention the epidemic of cholera in the 90s in Latin America that reduced the consumption of vegetables in many countries of the region for almost one year. Another example may be the two grapes with dangerous residues detected in the 80s in an entrance port of the USA, which severely affected Chilean exports. Also about that time, the Alar (daminozide, a growth regulator used to control fruit set) scare considerably diminished the consumption of apples in the United States.

Food safety is the absence of substances dangerous for health and particularly in fruits and vegetables, the presence of pesticide residues on the product has been the main concern for consumers. However, there are many other contaminants potentially as dangerous such as the presence of pathogenic microorganisms, mycotoxins, heavy metals, and others.

As fruits and vegetables are consumed fresh and are many times not peeled, all organisms pathogenic to humans which are carried on their surface constitute a potential danger. Bacteria, like Shigella spp., Salmonella spp., Aeromonas spp., Escherichia coli, Listeria monocytogenes and the toxins produced by Clostridium botulinum and others, have been identified as responsible for illnesses associated with the consumption of fruits and vegetables. The Hepatitis A virus has been detected on produce as well as parasites like Entomoeba histolyca, and Giardia lamblia.

Agrochemicals are one of the tools that man has used to satisfy the growing need for food. They are the herbicides, insecticides, fungicides, fumigants, rodenticides, growth regulators, waxes, disinfectants, additives and all other products of a chemical nature used during production or post-harvest handling. Their residues have always been a major concern, although the advances in chemistry and biochemistry, the better understanding of ecology as well as pests and diseases, and the growing use of non-chemical control methods, have made the present world a much safer place. Today's agrochemicals are less toxic and persistent, their degradation products are generally innocuous and better laboratory methods have been developed for their detection. Additionally, there is a greater consciousness about their use, waiting times, doses, etc.

Each country has its own legislation in terms of the Maximum Residue Levels (MRLs) within the framework of the Codex Alimentarius Commission or other international organisations. An MRLor tolerance is the maximum concentration of pesticide residue allowed resulting from its application according to correct agricultural practices.

Agrochemical use should be within the Good Agricultural Practices framework to guarantee maximum safety and to minimise risks to the consumer's health. Specific products should be used to control pests or diseases following the manufacturer's indications, particularly those refering to crops on which they can be used, as well as minimum waiting times between application and harvest.

Other health hazards are the presence of nitrates in leafy vegetables, oxalates in some species and heavy metals accumulation, particularly when sewage is used as fertilisers or organic amendments. Some toxicity may exist in some natural compounds produced by the crop itself or by the fungi that colonise its surface like the micotoxins.

Producing Quality Product

Producing a quality product begins well before planting the seed. Soil selection and preparation, its fertility and irrigation aptitude, weed control and crop rotations, variety selection and other decisions have an influence on the quality of the product. In the same way, quality is affected by the climatic conditions during the growing period, as well as irrigation, fertilisations, control of pest and diseases and other cultural practices. Harvest is the end of cultivation and the beginning of post-harvest actions during which preparation for the market, distribution, and sale take place.

Fruits and vegetables are highly perishable products and before being detached from the mother plant all demand water and nutrients. Once harvested, however, they depend on their reserves to continue living. Respiration, transpiration and the continuous changes taking place determine the internal and external quality. Deterioration rates depend on the type of product, growing conditions and other factors, but mainly on the conditions in which the produce is maintained after harvest such as temperature, relative humidity, movement and composition of the air, etc. Post harvest changes can only be delayed within certain limits and thus preparation for the fresh market should be quick and efficiently performed to avoid quality losses.

Post-harvest losses due to microorganisms can be severe, particularly in warm climates with a high relative humidity. Rotting produce contaminates the rest and under these conditions ethylene production is stimulated accelerating the rate of deterioration.

Most of the fungi and bacteria that attack fruits and vegetables after harvest, are weak pathogens and they mainly invade tissues through wounds. Injuries produced during handling provide numerous entrance routes to these pathogens although some of them are able to invade healthy products. Unripe fruits are usually more resistant to pathogen attack. It is also possible that infection occurring at the immature stage shows up later, when the natural defences are weakened by the ripening process.

A good disease control programme at field level reduces the source of inoculum and the risks of infections after harvest facilitating the control of post harvest diseases. Also, careful handling during harvest and packing operations reduces the physical damage that facilitates the establishment of microorganisms. Controlling temperatures to which produce is exposed is one of the main tools to control post harvest diseases since it diminishes the metabolic activity of the microorganisms and, by reducing the rate of the ripening process, the natural defences of the product are kept high. Controlling the relative humidity, particularly to avoid the condensation of water on the product, as well as controlled atmospheres is also useful in the control of post-harvest diseases.

Total Quality in Fruits and Vegetables

The concept of quality as a way to differentiate products has been recognised for years. As local or regional trade internationalises, quality consolidates as the main competitive tool for excellence, reinforcing the need to establish standards to separate quality into categories or degrees, as well as to define the limits of allowed defects. Nowadays, domestic and international trade of fruits and vegetables is regulated by quality standards in most countries, providing a common language among the different participants of the production-commercialisation-consumption chain.

Standards are also the legal framework to settle commercial disputes and are useful as a basis for reporting on market prices as prices only can be compared between the same quality category. The quality system established by the standards is known as "Inspection for quality" where representative samples at the final stage of preparation for the market should fulfill the specified limits and their tolerances. Although it is easy to apply, it has, at least, two big disadvantages: firstly, they are not totally adapted to highly perishable products where quality varies continually. Secondly, its application does not improve the quality of the product, it only separates in degrees the quality that comes from the field.

At the same time that quality standards developed and were applied, new ideas began to be conceived by industry. Firstly, it became evident that a systematic and preventive approach was much more effective and economic to improve quality than the elimination of the faulty units at the end of the line, when the production and packaging costs had already happened. Secondly, it became clear that the quality concept extends beyond the product itself, since it is affected by the systems and procedures involved in the production and preparation for the market. Finally, the consumer's opinion begins to be more and more important. It is no longer sufficient for a product to be technically perfect and produced in an economically profitable way, it is also necessary to satisfy the consumers'expectations of quality.

The application of statistics to control the variability of the different units in the production lines gave birth to the system called "Quality control" or "Statistical Quality Control", which was adopted by most manufacturing companies in the first half of the 20th century. This method or system essentially provides the analytic tools for monitoring the production process and for taking measures when variability exceeds certain limits considered as normal. Its application improves the quality of the process contributing greatly to improve the quality of the product. They are tools that can be applied at the fruit and vegetables packinghouse level.

This system was transferred to Japan after World War II where it evolved into what today is known as "Total Quality Management" or simply "Total Quality". Total quality is today the most complete conceptual framework to assure quality to which each person or activity within the production process is committed, aiming at zero defects and customer's complete satisfaction, even going beyond his/her expectations. At the same time that TQM was developed, the concept of "Quality Assurance" was coined in Europe.

Its scope is slightly narrower than TQM, but a lot easier to implement and probably better adapted to fruit and vegetables. It is defined as all those planned and systematised actions necessary to guarantee that the product or service will satisfy the requirements of quality. It usually requires the fulfilment of certain rules, protocols, or standards developed specifically and with a certification by an independent company authorised to grant it. The ISO system is probably the best known and within it the series 9000. It is also appropriate to mention the HACCP (Hazard Analysis Critical Control Points) system, designed specifically to guarantee that food is not exposed to any type of contamination that could put health at risk. Today, this method is recognised internationally as the logical and scientific tool for all food quality systems. It is also preventive in nature and the key element is the identification of the critical points, within the process, where quality should be controlled to prevent, eliminate or reduce to acceptable levels all possible safety risks.

The HACCP system is required today in the USA and other countries to import meat, fish, eggs, and other foods. Up to now, it is not required for fruits and vegetables, although different export countries are already implementing it to assure a superior quality of their products. The logic of the HACCP can be applied to the detection of other defects of quality.

Although all these systems have their origin in industry, their application extends to other sectors. Agriculture, and particularly the production of fruits and vegetables, is now incorporating many of the methods and ideas conceived by the industrial sector because the basic principles are not only applicable but also recommendable for highly perishable products where quality deteriorates quickly.

Several export companies have implemented the HACCP together with the ISO 9002 certification, which guarantees food safety within a system of quality assurance. Akey concept is that quality systems are not mutually exclusive but rather they overlap widening the application approach, extending beyond the product itself and embracing the preparation process, inputs, suppliers and intermediaries, besides the incorporation of the feedback from the client or consumer for its continuous improvement.

The basic principles of total quality can be summarised in the following way:

— The consumer is always first
— Each operation is part of a process
— Quality improvement never ends
— Quality is made, not controlled
— Prevention of quality problems is made through planning.
— The desired product should be obtained at the desired moment. Post harvest handling should be appropriate to reach the desired market under the desired conditions.

Microbiological Risk in the Production and Distribution

Products go through different stages of operations after harvest. This provides many opportunities for contamination besides those which naturally occur in the field. Consumers strongly reject foreign materials on products or inside packages. These include for, example, dirt, animal feces, grease or lubricating oils, human hairs, insects, plant debris, etc. However, because this is usually due to insufficient care in handling, they are relatively easy to detect and to eliminate.

A more serious problem is the presence of human pathogens on produce. These may not be visible or detected because of changes in appearance, flavour, colour or other external characteristics. It has been shown that specific pathogens are able to survive on produce sufficiently long to constitute a threat. In fact, many cases of illness related to consumption of produce have been reported

Three types of organisms can be transported on fruits and vegetables, which may constitute a risk to human health: virus (hepatitis A, for example), bacteria (*Salmonella* spp., *Escherichia coli*, *Shigella* spp., and others) and parasites (*Giardia* spp., for example). Mycotoxins and fungi do not usually constitute a problem. This is because fungi development is usually detected and eliminated well before the formation of mycotoxins. In most cases, bacteria is responsible for illnesses related to the consumption of fruits and vegetables.

Produce can become contaminated through mechanisms that are complex. The best strategy to obtain a safe product is to prevent contamination at various points throughout the production and distribution chain. This includes undertaking certain sanitary treatments and maintaining produce under conditions (mainly temperature) which do not favour the development of micro-organisms. This approach is known as the "systems approach". Every step of the process is a part of an integrated system.

Records and/or documentation of all activities and treatments are required so that a tracing scheme is established. In this way, it is possible to identify areas of weaknesses within the system and to take corrective measures. Following strict written procedures for good agricultural practices (GAP) and/or good manufacturing practices (GMP) are critical factors for the implementation of such a system. This should be combined with such systems as Hazard Analysis Critical Control Point (HACCP) analysis in order to identify the critical points at which known safety food hazards must be controlled.

Before Harvest

Some human pathogens are naturally present in the environment. However, fecal deposits(human, animal, or wild animals) are the main source of contamination of produce. Entry is mainly through irrigation or washing water. Microorganisms in surface water (rivers, lakes, etc.) may come from upstream dumping of untreated municipal wastes. Underground water may also be contaminated from septic tanks leaching through soil into aquifers. If only contaminated water is available, underground drip irrigation is the only irrigation system recommended to avoid contamination of above-ground edible plant.

The main causes of contamination include: animal manure or sewage waste used as organic fertilisers or the presence of animals in production areas. Manure should be composted aerobically to reach 60-80°C for a minimum of 15 days. Composting of static piles and earthworms do not guarantee that microorganisms have been inactivated. Wastewater and municipal wastes should only be used if effective disinfecting systems are available.

The production and harvest of fruits and vegetables tend to rely heavily on human labour. Other sources of contamination include the hygiene conditions of field workers. First, production fields are usually a long distance from bathrooms, together with other sanitary facilities for the hygiene of personnel, Second, hired teams of migrant labour temporarily live in the fields. Here, conditions and sanitary practices are considered unacceptable. In addition to providing portable toilets, staff need to understand the importance of proper hygiene practices for food safety.

The type of product also has an influence: in vegetables which are characterised by low acidity in tissues, bacteria tends to dominate, while fruits are mainly colonised by fungi. Crops growing close to the ground like strawberry and leafy vegetables in general, are more susceptible to contamination by water, soil, or animals when compared to tree crops. Finally, some chemical constituents of tissues such as organic acids, essential oils, pigments, phytoalexins etc, have antagonistic effects and provide some form of protection against the development of micro-organisms.

As with other handling operations, numerous opportunities exist for contamination during harvest. Wounds and bruises may exude latex and other plant liquids from tissues and provide the substrate for the growth of microorganisms transmitted by hands, tools, clothes, water, or containers. Contamination at any point throughout the distribution chain can be exacerbated by the conditions to which the produce is exposed. Temperature is the most important factor to be considered.

Preparation of Product for Market

In packinghouses or processing plants, people who are ill or have open wounds should be forbidden from making contact with the product. Workers should use hairnets and clean outfits when handling the product.

Street clothes must be left outside the working area and eating or drinking in the pack-house should be forbidden. Workers should wash their hands daily at the beginning of operations and every time they return to work, particularly after having used restrooms.

However, the main source of contamination in terms of product preparation for market, is probably water. Water is essential for packinghouse operations either for washing the product, containers, and facilities or for dumping, hydrocooling. Other uses include personal hygiene or as a medium for waxes, chemicals, etc.

Water disinfection

Water impurities are frequently concerned with: the suspension of materials, microorganisms, organic matter, off-colour, off-odour as well as minerals and dissolved gases. Municipal water is filtered and treated (normally with low chlorine concentrations) to ensure that it meets the chemical and microbiological requirements for food safety and contact with food. Alternative sources of water must be filtered and sanitised.

Sanitation is necessary to avoid the spread and contamination to other units, even with the use of municipal water. Different methods exist for water disinfection. These include chemical, thermal, ultrasonic waves or irradiation. In postharvest operations, chlorine and its derivatives are the cheapest and most widely used substances to destroy bacteria and fungi in water as well as on the surface of fruits.

Chlorine is a gas with a strong and penetrating odour and is extremely reactive chemically. At post harvest level, it is used mainly in three different forms: as pressurised gas from metal cylinders, as calcium hypochlorite (solid) or liquid as sodium hypochlorite, commonly known as "bleach" for household whitening and sanitising. Chlorine gas is difficult and dangerous to handle and is normally limited to large operations such as municipal water treatment. Calcium hypochlorite is widely used in concentrations of 65% but is difficult to dissolve in cold water. Sodium hypochlorite is more expensive than the other two formulations in terms of chlorine concentration (5 to 15%). However, its easy dosage makes it convenient for small size operations.

In aqueous solution, chlorine exists as hypochlorous acid, hypochlorite ion, or as a variation of both, depending on the pH solution: the former predominates in acidic solutions while the latter in alkaline solutions. Germicidal action of hypochlorous acid is about 50-80 times higher than hypochlorite. In order to maximise the effect on microorganisms, the pH solution should have a range of between 6.5 and 7.5. Below this range, the hypochlorous form is extremely unstable and tends to escape as a gas resulting in irritation and discomfort to workers.

It also becomes extremely corrosive for the equipment. On the other hand, its effectiveness as a sanitiser is significantly reduced above 7.5. To keep pH values within permitted ranges, vinegar could be used to acidify while sodium hydroxide could be used to alkalinise. Maintenance kits for swimming pools could be used to monitor pH. Chlorine as a gas lowers the pH of the solution while as hypochlorite, either of calcium or sodium, this will be increased.

Concentration of active chlorine is expressed in parts per million (ppm). Concentrations of active chlorine in the range of 0,2 to 5 ppm are able to kill most bacteria and fungi present in water. However, in commercial operations higher concentrations are used (100-200 ppm) for washing and hydrocooling. Aliter of household bleach (80 g active chlorine/dm^3) in 400 liters of water represents about 200 ppm, and in 800 and 1600 liters, about 100 and 50 ppm, respectively. It is convenient to start daily operations with: low concentrations (100-150 ppm); to increase the amount of chlorine in solution as water becomes contaminated with dirt and plant debris and an increase in microorganisms.

Quick exposure (around 3-5 minutes) is adequate for the purposes of disinfection. However, in addition to pH level and impurities, the temperature of the solution is also important. This is because low temperatures can reduce activity. Another important factor is the extent to which microorganisms can be developed. This is because their spores are from 10 to 1000 times more difficult to kill compared to their vegetative state.

The use of chlorine in fruits and vegetables is banned in some countries. This is because it is possible for it to react with organic matter leading to the formation of chlorate compounds and trihalomethans. These are suspected of being carcinogenic. As a result of this, the industry is investigating alternative sources of sanitisers.

Ozone is a gas with a strong oxidizing action in concentrations of 0,5-2 ppm. It has been approved for water sanitation. However, it is difficult to apply. This is because reliable methods for monitoring the level of concentrations do not currently exist. In addition to this, it is only effective within a reduced pH range of (6-8) and must be generated in the same place of application. It is dangerous to humans in concentrations higher than 4 ppm and may cause damage to some plant tissues. In spite of these limitations, it is probably the most promising compound to replace chlorine. Ultraviolet light in wavelengths of 250-275 nm can also be used. It is unaffected by water temperature or pH. However, water must be filtered as turbidity reduces its efficacy.

Water management is also important. This is because several washes are required. This is more effective than a single wash. The following is an outline of a procedure for good cleaning practices: First, undertake an initial wash to eliminate dirt and plant debris. Second, wash with chlorinated water and finally rinse with plain water. Brushing or water agitation increase washing efficiency. Water recirculation should be undertaken in reverse to the flow of the product, i.e. rinsing waters can be reused for the initial washing. Hydro cooling is one of the most efficient systems for precooling. It is, however, also one of the most risky in terms of microbial contamination. This can be caused by water infiltration inside the fruits. As a result of this, it is important to consider alternative precooling methods such as forced air.

Plant hygiene

Industrial facilities follow a system of strict hygiene measures. However, when produce is prepared for the fresh market, limited attention is normally paid to hygiene facilities. This is particularly the case when low cost materials have been used in the construction of the packing shed.

Although other factors need to be taken into account prior to layout and organisation, it is important that the packing shed be designed to allow for thorough cleaning procedures. The reception area should be kept separate from the delivery area. Similarly, "clean areas" or areas where product is prepared, should be kept separate from other areas or where product is handled as receivables from the field. There should be a clean area where workers can take breaks, change clothes, and take care of personal hygiene. For example, the availability of hot water, showers, and clean restrooms within a clean and comfortable environment.

While it is a requirement that dust and other impurities are eliminated, liquid sanitisers should be used to disinfect the facility and equipment, particularly those in contact with produce. Chlorine based sanitisers are the most widely used disinfectants. However, the choice depends on type of water, pH, cost and the type of equipment. Iodine based disinfectants (iodophors) are less corrosive to metals than chlorine. They are unaffected by organic matter but their effectiveness within the pH range is quite narrow (2,5-3,5). In addition to this, they may stain surfaces. Quaternary ammonium compounds are widely used for disinfecting floors, walls, and aluminum equipment. While effective over a wide pH range, they are not affected by organic matter and are non-corrosive. However, they are expensive and leave residues on surfaces. There are other sanitisers available on the market that can be used in food plants.

Animals of all types, including mammals, birds, reptiles, and insects can spread microorganisms with their droppings. Their entry into packing areas should be forbidden, including pets. It is necessary to seal crevices and keep doors, windows, and air inlets closed or fitted with insect-proof screens. It is also important to have insect and rodent control programs, together with approved pesticides, traps, and baits. Facilities and the surrounding environment must be kept clean and tidy to prevent the nurturing and harbouring of insects, rodents, reptiles and other animals. Garbage and waste materials should be removed daily.

Storage and transport

In terms of sanitation, there are two potential sources of risk: contamination by human pathogens and their growth and development due to the favourable conditions created because of the packaging environment or storage area. Factors highlighted earlier about hygiene of personnel and facilities are also applicable here. Two additional recommendations include: the use of new containers and not undertaking repackaging. Other considerations to prevent cross contamination include not storing or transporting fruits and vegetables with other fresh food items.

The best strategy to prevent growth and development of human pathogens is to keep product at recommended storage conditions, particularly temperature. Microorganisms can be divided into three main categories according to their level of adaptation to temperature. These are: psychrotrophic - the ability to grow under refrigerated conditions, although the optimum ambient temperature is (20-30°C); b) mesophilic - those that develop best at ambient temperature (20-40°C) but not under refrigeration and c) ther-mophillic - requiring temperatures above 40°C. The last two (i.e. b and c) do not concern fruits and vegetables for the fresh market, but may be present in inadequatelt processed items. Generally, refrigeration inhibits growth of microorganisms, but psy-chrotrophics may develop on produce if storage time is too long.

The atmosphere at which a product is stored also influences microbial development. *Clostridium botulinum*, for example, is not a concern when a product is prepared for the fresh market. However, it may develop and produce toxins on tissues with a pH higher than 4,6 and under conditions of low oxygen. It may be present in inadequately pasteurised canned products but it may also develop under modified atmosphere conditions. It has been reported that this bacterium can cause human poisoning.

Sale

Fruits and vegetables may become contaminated at the point of sale, storage, and home preparation. The previous discussions on personal hygiene and prevention of contact with animals are also valid here. The common practice at retail of cutting large fruits into portions (pumpkins, watermelons, melons, etc.) should be avoided and refrigeration is recommended for the most perishable items.

Fruits and vegetables are microbiologically safer compared to meat, milk, poultry, and other foods. However, during the process of fighting organisms they do not usually undergo a kill step so many organisms are not destroyed (by cooking) and are therefore potentially dangerous if contamination exists. It is difficult to estimate how real this potential threat is. This is because it is usually reported when it becomes serious. In addition as fruits and vegetables are regarded as "healthy food", they are not considered to be the cause of food-related illnesses.

Instead, other food eaten during the course of the same day is usually to blame. However, available evidence seems to indicate that this is increasingly becoming a problem. The reasons for this may be twofold: First, there is a trend towards environmentally friendly practices in agriculture. The use of manure based organic fertilisers or soil amendments increases the risk of contamination. Second, the concentration of supply particularly through supermarket distribution centers supplying a large number of stores, means that a single case of contamination can have an enormous impact throughout the system.

To obtain high quality produce with minimum risk levels, the first step is to understand the complexity of microbial contamination and recognise its importance. The examples of good agricultural and manufacturing practices described in these paragraphs may not apply to all fruits and vegetables. However, they may be useful to devise specific preventative measures. With the current level of technology, it is not possible to eliminate this risk.

However, it is important to know how to minimise it as much as possible. It is cheaper and more effective to prevent microbial contamination of fruits and vegetables than to have to confront the problem when it occurs. A successful food safety program requires serious commitment from everyone throughout the food chain from production to consumption. Key factors to be considered are: the availability of trained personnel, and a system that ensures that there are no missing links in the quality chain as regards contamination prevention.

Many different testing procedures exist for the detection of microorganisms, such as total plate counts or aerobic plate count. They give an idea of the extent of microbial contamination, but have limited value in terms of assessing food safety. A wide range of microorganisms exist naturally on the surfaces of fruits and vegetables and they will colonise a growing medium. However, this does not mean that they are a health hazard. These types of tests are useful for monitoring the hygiene system or evaluating the impact of certain sanitary measures.

Detection of *Salmonella* spp., fecal coliforms, *E. coli*, and other pathogens require specific tests and their lack of detection does not mean that the produce is free from other harmful microorganisms. As a result, the best strategy is to minimise the risk and to prevent as much contamination as possible. An important aspect in any program of good agricultural or manufacturing practices is to have a tracing system. This is because it is possible to identify and pinpoint quickly problems of contamination. The necessary corrective measures can be undertaken as soon as possible. The short lead time between harvest and consumption of fruits and vegetables makes it difficult to react in time if an outbreak is detected.

Refeences

Hardenburg, R.E., A.E. Watada, and C.Y Wang. 1986. The commercial storage of fruits, vegetables, and florist and nursery stocks. *USDA Agricultural Handbook* 66.

Kader, A.A., D. Zagory, and E.L. Kerbel. 1989. Modified atmosphere packaging of fruits and vegetables. *Crit. Rev. Food Sci. Nutr.* 28:1-30.

Knee, M., ed. 2002. *Fruit quality and its biological basis.* Sheffield: Sheffield Academic Press, 320 pp.

Thompson, A.K. 1998. *Controlled atmosphere storage of fruits and vegetables.* Wallingford, UK: CAB International, 288 pp.

12

Food Loss Prevention in Perishable Crops

Post-harvest losses of fruit a and vegetables are more serious in devoloping oountries then those in well developed oountries. An additional constraint to improving this situation is that in moat developing countries the number of scientists concerned with post-harvest food losses is significantly lower than those involved in production research. In the early days of horticulture in wolf developed countries, heavy losses occurred in much the same manner as they do today in developing countries. Increasing industrialization in technologically advanced nations gradually brought about improvements in crop handling. Elaborate harvesting equipment replaced the crude harvesting tools. Collection centres were strategically established in major producing areas. Containers were remodelled to add more protection to the produce. Commercial storage plants were installed and grade standards adopted. Engineers and economists became more and more aware of raw material behaviour. Concomitant advances in Refrigeration Technology in the developed countries have made possible establishment of cold chains for the entire post-harvest and handling operations. At the institutional level post-harvest research was initiated. Pilot packing houses were installed, coupled with the development of intensive training programmes. the improvement of product quality and reduction in post-harvest losses became the main concern of producers, middlemen, marketing specialists and consumers. Today, enormous volumes of quality horticultural crops produced in technologically advanced countries are made available to millions of people through improved post-harvest handling. Thus, historically and by necessity, post-harvest technology is part of the normal development processes in agriculture.

These handling procedures are not fully recognized in less developed countries. Here agriculture may be characterized as disjointed. Production is not linked with marketing. With perishable crops like fruits and vegetables, storage, packaging, transport and handling technologies are practically non-oxiatant, Hence, considerable amount a of, produce are loaf. Thus, as more fresh fruits and vegatables are needed to supply the growing population in less developed countries, as more produce is transported to non-producing areas, and as more commodities are stored longer to obtain a year-round supply, post-harvest lose prevention technology measures become paramount. It is distressing to note that so much time is being devoted to the culture of the plant, so much money spent on irrigation, fertilization and crop protection measures only to be wasted about a week after harvest. It is, therefore, important that post-harvest procedures be given as much attention as production practices the stages from planting until the product a roach the consuming public must be a mutual undertaking between the growers and those who will handle the products after harvest.

In the warm humid tropics, fresh fruit a and vegetables have an extremely, low level of natural protection against the climate peat a and bio-chemical and physiological deterioration. It is also in the warm humid tropics that human diseases are moat numerous and widespread and consequently where the need for the nutritional value of fresh fruits and vegetables la moat essential. On the other hand, dried grains and legumes, and to a certain extent, root crops (yams), have a fair degree of natural protection against the various deteriorative elements, even though considerable post-harvest losses do occur.

Importance of Perishable Crops

In terms of importance as assessed by volume of production within the tropics these are cassava, potatoes, yams, sweet potatoes, taro and others. It is estimated that some 174 million tone of roots and tubers are produced annually of which cassava constitutes about 60%. Cassava is almost entirely produced and consumed in developing countries although in recent years there has developed a substantial trade in dehydrated cassava chips which are exported from developing countries to European countries as a low-goat animal feed ingredient, The European Economic Community imported 2.3

million tons of dried cassava chips in 1975. Next in volume of production are potatoes, yams and sweet potatoes. These make contributions respectively about 14, 11 and 10%. At the end of the scale taro contributes a lowly 2.2% while miscellaneous tubers account for 2.5%.

The total annual production of horticultural crops in developing countries is 620 million tone which is 85% of the cereal production in these countries. This surpasses the relative world average whore horticultural crop production is only 70% of the cereal production. These high production figures indicate the important role that horticultural crops play in the economy of developing countries although it must be remembered that these crops are far higher in water content than are the cereals.

Horticultural crops are essential for a nutritionally balanced diet. Fruits and vegetables are the major source of vitamins A and C, a good source of calcium and iron, and they supply part of the requirements for a number of other minor nutrients. Roots, tubers, bananas and plantains are important sources of calories and also supply a number of minor nutrients, and some protein.

In addition, horticultural crops add variety, enjoyment and a sence of satisfaction with the diet because of their appealing colours, flavours and textures. For example, it has been said that although onions and garlic are not rich in nutrients, they make a vegetarian diet acceptable because of the savoury flavour they impart to the monotonous starchy diet in a developing country. It is no accident that onions rank third, and garlic ranks tenth in volume of vegetable production in developing countries.

On all of these counts- economics, nutrition, acceptability- horticultural crops play a major role in developing countries, amply justifying the contortion that something should be done to reduce the high losses that presently occur in these commodities.

Sites of Losses

Losses may occur anywhere from the point where the food has been harvested or gathered up to the point of consumption. For the sake of convenience the losses can be broken down into the following sub-headings:

a. *Harvest.* The separation of the commodity from the plant that produced it. In the case of roots, tubers and bulbs the commodity is lifted out of the soil.

b. *Preparation.* The preliminary separation or extraction of the edible from the non-edible portion, e.g., the peeling of fruits and vegetables.

c. *Preservation* is the prevention of lose and spoilage of foods. For example, the sun-drying of fruit, the use of refrigeration and the use of funficides to inhibit mold growth in fruits.

d. *Processing* is the conversion of edible food into another form more acceptable or more convenient to the consumer, for example, the manufacture of fruit juice and the canning of fruits and vegetables.

e. *Storage* is the holding of foods until consumption. Most storage is common storage (ambient temperature) but there are extensive storage capacities that can hold food under refrigerated or controlled atmosphere conditions.

f. *Transportation.* All forms of transportation are used to convey foods from the point of production to the ultimate point of consumption.

Post-harvest losses in fresh root/tuber crops have their origin in mechanical damage, physiological processes, infection by decay organisms and, occasionally, pest infestation. The losses caused by these processes may occur during all stages of the food supply system from crop maturity, through harvesting, transportation and storage. The degree of lose associated with these factors is determined by the plant material involved, the prevailing environmental conditions and management of the food supply system.

Magnitude of Losses

Reliable statistics on losses are few. It is possible to find individual cases with losses ranging from 0% to 100%. The extent of losses is highly variable depending on a number of conditions. Stable foods such as cereal grains can be stored in good condition for several years, whereas perishable foods such as fruit a and vegetables, spoil quickly unless given special treatment such as canning and freezing. The longer the time the food is stored the more opportunity there is for losses of all kinds to occur. Perishable crops generally suffer from higher losses than the cereals.

A number of figures for the extent of loss is quoted in scientific literature and by the communications media, but much of this information is unreliable because the amount of loss has been estimated and has not been obtained by actual measurements. There is often the temptation to cite "worst case" figures to dramatize the problem.

Another problem is that even some of the figures that have been obtained by careful measurements are manipulated for various reasons. In some cases there is the temptation to exaggerate the figures of loss, particularly if there is a prospect that high figures of loss will prompt aid from donors. In other cases there is a temptation to minimize loss figures in order to prevent the embarrassment of acknowledging the magnitude of losses, or for political, financial or trading reasons.

Another precaution that needs to be taken in assessing overall losses is to ensure that the arithmetic of calculating loss figures is correct. The extent of losses can be unwittingly exaggerated unless the arithmetical calculations are correctly performed.

The pattern of losses varies widely from country to country. There is a marked contrast between the site of major losses in developed countries and developing countries. In a typical developed country losses may be fairly high during harvesting because the agricultural machinery that is used to harvest the crops leaves some of the commodity in the field and mechanically damages some of it. Considerable quantities of foods a may be discarded at the point of harvest because they are of the wrong size, shape or colour. These are planned losses. In developing countries harvest losses are usually lower because most of the crop is hand picked. The amount of material rejected in developing countries is less because the expectation of quality and uniformity is generally lower than for developed countries.

In developed countries losses are generally small during processing, storage and handling because of the efficiency of the equipment, good quality storage facilities, and close control of critical variables by a highly knowledgeable cadre of managers. In contrast, in developing countries losses in processing, storage and handling tend to be rather high because of poor facilities and frequently inadequate knowledge of methods to care for the food properly.

Loss Assessment

There appears to be no established generally accepted methodology for determining post-production lose in root/tuber crops. The National Academy of Sciences (USA) publication on post-harvest losses makes the following differentiation between assessment, measurement and estimation of losses:

— Assessment is a rough quantitative approximation of food loss or the characterization of the relative points of lose in a particular food supply system. This approach implies a measure of subjectivity resulting from a lack of sufficient information.

— Measurement on the other hand is a more precise quantitative observation with less subjectivity. With measurements there is a high expectation of reproducibility without observer bias.

— Estimation is the interpretation of a number of scientific measurements. Here the process of interpretation depends on the experience and judgement of the observer.

At whatever level of precision post-harvest lose is determined the value will be specific in time and for location. This is due to the fact that lose is a function of the condition of the material, the prevailing environment, the nature and intensity of bio-degenerating organisms and the crop material management. None of these are constant. They are all dynamic factors liable to continuous change. As a consequence, crop loss, however determined will always be variable. This is illustrated by yam storage. Some cultivars of the rotundata/cayenensis/discorea species remain dormant for about 3 months, during which period storage loss is low. When placed in the traditional yam barn the keeping characteristics of these yams are good. It is generally believed that little improvement can be achieved without radical change in storage technology. However where the tubers are infected by nematodes the storage potential is much reduced and high losses occur.

It would be useful to have a standard method for assessing losses for each type of commodity but this is a difficult task due to crop diversity, inherent perishability and the complexity of the marketing and distribution channels.

Effects of the Environment on Food Losses

The environmental conditions under which foods are stored and processed can have a major effect on the keeping quality of the foods and the amount that is lost. The major environmental influences on the keeping quality of foods are the following:

Temperature. In general, the higher the temperature the shorter the storage life of horticultural products and the greater the amount of lose within a given time, as most factors that destroy the produce or lower its quality occur at a faster rate as the temperature increases. For asparagus the loss of quality occurs at a more rapid rate than lettuce, reaching the point of being unsaleable in 3 days when stored at 25°C.

Humidity. There is movement of water vapour between a food and its surrounding atmosphere in the direction towards equilibrium water activity in the food and the atmosphere, A moist food will give up moisture to the air while a dry food will Absorb moisture from the air. Fresh horticultural products have a high moisture content and need to be stored under conditions of high relative humidity in order to prevent moisture loss and wilting, exceptions to this being onions and garlic. Dried or dehydrated products need to be stored under conditions of low relative humidity in order to avoid absorbing moisture to the point where mold growth occurs.

Solar Radiation. The solar radiation that falls upon foods held in direct sunlight increases the temperature above the ambient temperature. The amount of increase in temperature depends on the intensity of the radiation, the size and shape of the food' and the duration of exposure to the direct rays of the sun. The intensity of solar radiation depends upon latitude, altitude, season of the year, time of day, and degree of cloud cover. Under clear skies it is most intense when the sun is most directly overhead. Hence the intensity of radiation is greater in tropical zones than in temperate zones. As discussed above, a high temperature is deleterious to food quality and increases wastage. It is ironic to note that, in the temperate climates where the intensity of solar radiation is moderate, almost all food is kept inside under cover whereas in tropical climates, where solar radiation is much higher, considerable quantities of food can often be found in the direct rays of the sun deteriorating in quality at a rapid rate.

Altitude. Within a given latitude the prevailing temperature is dependent upon the elevation when other factors are equal. There is on the average a drop in temperature of 6.5°C for each Km increase in elevation above sea level. Storing food at high altitudes will therefore tend to increase the storage life and decrease the losses in food provided it is kept out of the direct rays of the sun.

Atmosphere. The normal atmosphere contains by volume, approximately 78% nitrogen, 21% oxygen, 1% argon, 0.03% carbon dioxide' various amounts of water vapour and traces of inert gases. Modifying the atmosphere can improve the shelf life and reduce wastage of certain foods.

One type of controlled atmosphere storage (CA) is refrigerated storage in which the level of oxygen is reduced to about 3% with the carbon dioxide content being raised to 1 to 5%, depending on the commodity. This CA storage may double the storage life over that of regular cold storage for certain varieties of apples and pears by slowing down the natural rate of respiration.

Many fruits, the "climacteric fruits", generate ethylene gas during ripening and the presence of this gas Accelerates the rate of ripening. If the ethylene is removed from the atmosphere surrounding these fruits as it is generated, their storage life may be extended. Experiments have shown that placing such fruits in a fairly gas-tight container with potassium permanganate, which absorbs ethylene gas, can substantially extend the storage life even at ambient temperature.

"Modified atmosphere storage" is another type of controlled atmosphere storage. This term denotes storage of horticultural products in a beneficial atmosphere other than air that is not under as close regulation as in CA storage. Modified atmosphere storage can be obtained in boxes of pears, apples, and cherries that are lined with polyethylene film which acts as a barrier to the escape of carbon dioxide and the ingress of oxygen. Another method of obtaining a modified atmosphere storage is by the addition of dry ice which increases the carbon dioxide in the atmosphere to some extent,

Time. The longer the time the food is stored the greater is the deterioration in quality an' the greater is the chance of damage and loss. Hence storage time is a critical factor in lose of foods, especially those that have a short natural shelf life. The time involved from harvest to consumption of perishable commodities is much shorter than that from planting to harvest. While production time could take about several years for fruit trees, and generally about three to four months for vegetables, the duration of post-harvest handling could be as abort as one day to a few weeks. Any improvement in post-harvest handling treatments would therefore involve less risk and would be more economical than improvements in production.

Biological pressures. Bacteria and fungi are always present in the atmosphere to contaminate food and cause spoilage should conditions favour their growth. However, it should be emphasized that the contamination or inoculation process with bacteria and fungi occurs to an equal extent during the harvesting process. Soil organisms as well as foliage pathogens can be introduced. Bacteria that cause disease in plants are not usually introduced from the air par so except in aerosols. Micro-organisms can multiply very rapidly whenever conditions are favourable for growth. The only foods that are free from micro-organisms are those that have been thermally processed, such as canned goods.

A similar situation occurs with insects. Insects are in the field and can accompany foods as they are brought from the field into storage. Most of the stored products' insects can increase is number by a factor of 10 to 50 times par month under favourable conditions. Stored food insects are ubiquitous, hiding in storage facilities and moving with stored foods when they are moved. Consequently we can assume that foods can become infested with insect pests at any time unless special precautions are taken.

Rodents and sometimes birds can exert biological pressures similar to those of insects and micro-organisms. The great capacity of living organisms for multiplication by geometric increase generates heavy biological pressures upon stored foods.

FACTORS RELATED TO THE POST-HARVEST SYSTEM

Technologies

The major technologies for reducing losses in horticultural products are listed below followed by a statement of probable environmental effects from the named procedure.

Gentle handling

Because of their soft texture. All horticultural products should be handled gently to minimize bruising and breaking of the skin. Bruising renders the product unsaleable to most people although it usually has minor effect upon the nutritional value. The skin of horticultural products is an effective barrier to most of the opportunistic bacteria and fungi that cause rotting of the tissues. Breaking of the skin also stimulates physiological deterioration and dehydration. Careful digging and movement of roots and tubers significantly reduces postharvest losses. Careful handling of fruits and vegetables to minimize bruising and breaking of the skin likewise is a well-known method of reducing postharvest losses as is the provision of adequate shipping containers to protect the produce from bruising' and puncturing of the skin. Reducing the number of times the commodity is handled reduces the extent of mechanical damage.

Environmental effects. There are no adverse environmental effects to this technology. Thus careful digging, harvesting and handling, and appropriate packaging end transportation are environmentally count methods for reducing losses. Also, since damaged skin is the major entry point for fungal infections, some of which produce mycotoxins, gentle handling can improve the safety of the produce.

Temperature control

It is well known that cooling horticultural produce extends storage life by reducing the rate of physiological change and retarding the growth of spoilage fungi and bacteria. Cooling is the foundation of quality protection.There are several ways of reducing the storage temperature of horticultural crops:

Cooling technique Environmental effect

a) *Keep out of direct rays of sun.* This is an easy low-cost method with minimal effect on the environment. Almost all societies can provide shade at low economic or environmental cost.

b) *Use natural cooling*, e.g., harvest during the cool early morning hours, open stores for ventilation during the cool of the night, utilize the cool temperature of high altitude or a natural source of cold water when available.

c) Evaporative cooling obtained by drawing dry air over a moist surface. Minimal environmental and economic costs. Restricted to areas of low humidity and low-cost water.

d) *Mechanical refrigeration.* Energy coats and economic costs are relatively high but give most positive control of temperature. Generated heat is dumped into the environment.

e) *Cool promptly after harvest.* High energy coat.

Since every degree reduction from ambient temperature increases storage life' every form of cooling is beneficial even if it is not optimum cooling, i.e.' simple low-cost cooling or refreshing the product is better than no cooling at all.

The optimum storage temperature for most temperate horticultural crops is close to 0°C. If they are cooled slightly below this temperature they freeze and suffer from "freezing injury" and spoil quickly. Most tropical horticulture crops however can be injured oven at temperatures above freezing point. This is called "chilling injury" and causes rapid deterioration in quality. The optimum storage temperature for most tropical horticulture crops is between 7 to 10°C; for yams and bananas it is about 15°C.

Although refrigerated storage is not often appropriate for some commodities such as yams, it should be considered an important element in the temperature management of a wide range of perishable crops because it gives the most positive and direct control of temperature. The popularity of refrigerated storage in some countries has suffered setbacks duo to occasional poor design of units and bad management. This has sometimes resulted in the impression that refrigerated storage is costly and unsuited for use in developing countries which is not necessary the case. Good cosign and proper management are essential ingredients in considering the introduction of refrigeration techniques as are the supporting infrastructures available within the post-harvest system. When studied on a case-by-case basis it seems likely that refrigerated storage will continue to find many successful applications to the needs of developing countries.

There are tour basic principles which must be correctly applied for successful refrigeration of perishable crops:

i. *Select only healthy products.* Refrigeration does not destroy pathogens responsible for product deterioration but only slows their activity; it does not improve product quality but only maintains it. A damaged product will deteriorate more quickly than a healthy one even in refrigerated storage, hence it is pointless to submit poor quality produce to refrigeration. In addition storage under refrigeration increases the coat to the product. The storage therefore of low grade diseased produce frequently cannot achieve an adequate economic return.

ii. *Timely cooling:* Since refrigeration slows the development of micro-organisms and physiological changes responsible for deterioration of perishable crops, it is obvious that cooling should be applied as soon as possible after harvest. The technique of pre-cooling was developed to fill this need by cooling produce soon after harvesting down to a temperature appropriate to that product.

iii. *Adhere closely to optimal conditions for temperature and relative humidity.* It is well known that refrigeration provides maximum storage life it these two parameters are correctly adhered to. This fact is especially important for tropical fruits and vegetables because their optimum storage temperatures vary considerably between varieties and even between producing areas. One of the main roles of research centres in tropical countries should be to define optimum storage conditions for commodities grown under tropical conditions. There is a need to evaluate the limitations of storage of these commodities under a range of temperature conditions and to consider the implications and problems of product compatibility under conditions of mixed commodity storage.

iv. *Uninterrupted cooling*: Refrigeration should be applied from the point of harvest through to the point of consumption where maximum post-harvest life with high product quality is justified. This is the well known concept of the "cold chain".

High humidity

High humidity retards wilting and maintains the product in better condition. Most horticultural products store best in an atmosphere that has a relative humidity of 90%. Providing humidity has little environmental cost.

Waxing of the surface

Waxing the surface of horticultural products is a treatment used on a number of commodities including citrus fruits, apples, rutabagas and cucumbers. It retards the rate of moisture loss, and maintains turgor and plumpness and may modify the internal atmosphere of the commodity, and is performed primarily for its cosmetic effect; the wax imparts a gloss to the skin and gives the produce a more shiny appearance than the unwaxed commodity. Sometimes antiwaxing is a techniques that could probably be used more widely in developing countries with advantage. In some countries indigenous waxes may be suitable for this purpose. For example, experiments in Colombia have shown that waxing of cassava can extend the storage life from 2 to 3 days up to about 30 days by preventing discolouration in the vascular tissue. Work in India has also demonstrated the efficacy of indigenously produced wax emulsion formulations in extending the storage life of different fruits and vegetables.

Environmental effects. The waxes and wax formulations that are used in the U.S. are approved by the Food and Drug Administration and are kept under continuous review. Most of the ingredients in the wax mixtures are classified "Generally Recognized As Safe" (GRAS). In most cases the skin is removed and discarded before consumption in which case the wax is not ingested and should cause no special problems. However, problems might arise if unregistered formulations are used, or if the skin is eaten by humans or fed to animals.

Controlled atmosphere storage

Controlled atmosphere storage consists of placing a commodity is a gas-tight refrigerated chamber and allowing the natural respiration of the fruit to decrease the oxygen and increase the carbon dioxide content of the atmosphere in the chamber. Typically, for storage of apples the oxygen content is lowered to about 3% and carbon dioxide is allowed to increase to 1 to 5%. This atmosphere can extend the storage life of apples by several months and allows fresh apples to be marketed every month of the year. This technology requires expensive storage chambers and close supervision of the composition of the atmosphere and is unsuited for widespread use in less developed countries.

Some roots and tubers are stored in pits in the ground, known as "clamp storage". Well designed clamps tend to change the atmosphere to some extent by reducing oxygen and increasing the carbon dioxide content. Modified atmosphere storage would probably be effective for a limited number of commodities in developing countries especially if coupled with low temperature storage.

Environmental effects. Since this technology only manipulates the proportions of asses that are naturally present in the air there should be no adverse environmental effect. The now technology of hypobaric storage is emerging which maintains reduced pressure in the refrigerated storage chamber by means of vacuum pumps. In this system the commodity is placed in a flowing stream of highly humidified air which is maintained at a reduced pressure and controlled temperature. Under these conditions, Bases released by the commodity that limits its storage life, are flushed away. Reports indicate that the storage life of certain fruits and vegetables is extended substantially by this procedure. The economic feasibility of this type of controlled atmosphere storage is presently being tested. This is an energy-intensive and capital-intensive technology and is perhaps unsuited for less developed countries. The major environmental effect is the high energy coat.

Field factors

Maturity at time of harvest is an important factor in the keeping quality of horticultural products. Commodities that are harvested in an immature state not only have poor eating quality but may tend to shrivel in storage and be more susceptible to storage disorders. When picked too mature the commodity is soft or fibrous, the flesh breaks down more quickly and it has a shorter storage life. There is an optimum time of harvest to give maximum storage life for fruits, vegetables and tubers.

Suberization and curing

Potatoes, sweet potatoes, yams and several other roots and vegetables have the ability to heal skin wound e when held at moderately warm conditions end high humidity for several days after harvest. The self-healing of wounds, cute and bruises is known as curing. There are two steps in the curing process. First is suberization - the production of suberin and its deposition in cell walls. The second is the formation of a cork cambium and production of cork tissue in the bruised area. The new cork tissue seals the cut or bruised areas and helps prevent the entrance of decay organisms. The healing of injuries received in harvesting and handling prolongs the storage time and reduces the incidence and spread of decay in storage.

The storage life of onions and garlic is extend-d by exposure to warm dry conditions for several days to dry the outside akin and prevent the ingress of spoilage organisms. This process if also known as curing although physiologically it is rather different and causes about 5% weight loss. Curing is carried out in the field when weather conditions are suitable; otherwise the product is subjected to forced circulation of warm dry air when first put into storage.

Genetic control of shelf life

Each variety of a horticultural crop has a limited storage life even under optimum storage conditions. The potential storage life is partly under genetic control and can be manipulated by breeding.

Plant breeders should be encouraged to include potential storage life as one criterion in their programme for breeding improved varieties of roots, tubers, fruits and vegetables. This is particularly needed with the breeding programmes in tropical climates where refrigerated storage capacity is in short supply. This should be a high priority method for reducing losses in horticultural products.

Farmers should be encouraged to grow varieties that hare long storage life. For example, Martin and Degras point out that different Jam varieties differ in storability from a week to several months. Extension agents and experiment stations should be encouraged to include inherent storage life as one of the considerations to be taken into account when deciding which types of crops and which varieties of those crops should be recommended to farmers. There are no known adverse environmental effects from the efforts of plant breeders to extend the inherent storage life of horticultural crops. However, the results of plant breeders' work may need to be monitored. The U.S. Food and Drug Administration established regulations for release of new varieties of edible plans a when it wee discovered that a new potato variety that wee released in 1969 had an unusually high content of the toxic glycoalkaloids that are naturally present in potatoes. The FDA regulations apply to any plant material that provides more than 2% of the U.S. diet. The regulations require that plant breeders must establish ah two point a before releasing a new variety:

(i) that the content of the major nutrients is no lower than the average found in existing varieties of that commodity and (ii) that toxic substances naturally present in the commodity are no higher than normal for existing varieties.

Shorten the time between harvest and consumption

In developing countries a considerable amount of produce is wasted because of poor transportation systems and poor marketing procedures. Much produce is spoiled because it is stored beyond its inherent shelf life before marketing is completed.

Improving transportation and marketing facilities, spreading the harvest season by growing varieties that mature at different times, and staggering the planting dates of annuals and reducing the number of steps between producer and consumer are methods that can be used to shorten the time between harvest and consumption.

Processing

Considerable quantities of fruit a and vegetables are processed by dehydration, canning and freezing in developed countries. In developing countries. In developing countries small amounts of these commodities are processed for local consumption although large volumes of some commodities are processed for export.

Canning and freezing require a high capital cost, high energy costs and expensive packaging and are unsuited for widespread use in less developed countries. Dehydration or sun drying is the simplest and lowest cost method of preservation and should be more widely used in developing countries because it converts a perishable commodity into a stable item with long storage life. Some excellent quality dehydrated products can be made from roots and tubers; this kind of processing should be encouraged.

Environmental effects. Occupational hazards in the fruit and vegetable processing industry are the normal hazards associated with machinery, for which adequate safety measures are well developed. The National Institute for Occupational Safety and Health in the U.S. (NIOSH) have no complaints en safety hazards in processing plants that handle horticultural products. The U.S. Occupational Safety and Health Administration (OSHA) have no regulations specific to the fruit and vegetable processing industry other than the board guidelines that apply to industry in general. The fruit and vegetable processing industry is not on the list of occupational groups in which excess cancer incidence is reported by the U.S. National Cancer Institute.

There are occupational risks to some workers with specific horticultural products. For example, Barber and Husting report isolated cases of contact dermatitis among workers handling raw fruits and vegetables, including carrots, asparagus, mangoes, cashew fruits and nuts, and some citrus fruits. Fruit and vegetable handlers may also suffer contact dermatitis due to sensitivity to specific insecticides and fungicides. Indirect effects of handling fruits and vegetables include chapping and moniliasis from exposure to moisture, photosensitization dermatitis from sunlight, and parasitism from mites. Products that cause photosensitization include fig, rue, lime, bergamot, paranips, parsley, carrots, fennel, dill and pink rot celery. Raw pineapple fruits contain the proteolytic enzyme bromelain which causes skin irritations to workers in pineapple processing plants. This problem is overcome by supplying workers who handle cut fruit with rubber gloves.

Heat treatment

Some of the organisms that cause rotting are inhibited or killed at elevated temperatures that are below the injury threshold of the product. For example, hot water dipping of mangoes at about 50°C for a few minutes kills many pathogens without adversely affecting the quality of mango. Heat treatment is however not a desirable procedure for most fruits and vegetables. When applicable, very rigid temperature controls are needed. There is little adverse environmental effect from heat treatment. Small amounts of heat are dumped into the environment.

Sanitation

All handling, storage, cleaning and washing equipment for horticultural products should be kept in a sanitary condition in order to minimize the risk of spreading infection. Diseased or damaged units should be sorted out and properly disposed of because their presence promotes the growth of fungi and bacteria. Insects infesting cull piles may fly to good produce and introduce pathogenic organisms and increase losses. Wash water should be changed at regular intervals before it becomes heavily contaminated with fungi and bacteria and spreads infection. In some cases the wash water is treated with chlorine or some other chemical in order to reduce the count of viable organisms. The sanitation programme in the People's Republic of China is considered an exceptionally important element of pest control.

Use of chemicals

A number of chemicals may be applied to horticultural products in order to obtain a desirable post-harvest effect. Most of these are applied after harvest, but a few are applied in the field in order to obtain a specific post-harvest response. For example, the sprouting of onions in storage can be delayed by spraying the onions with maleic hydrazide (MH) in the field while the tops are still green. Chemicals used pre harvest whose scale propose is to achieve a post-harvest effect should be included in the list of post-harvest chemical treatments.

Post-harvest chemicals are classified into groups below. Many of these are not used commercially and are of research interest only:

a. Fungicides which prevent or delay the appearance of rota and molds in the product. Examples are, sodium orthophenylphenate (SOPP), benomyl, thiabendazole (TBZ), sodium hypochlorite, and sulphur dioxide (SO_2). Methyl formate (Erinol), ethyl formate and (in some countries) ethylene oxide are frequently applied to dried fruits to kill infestations of insects and molds. Sulphur dioxide and benzoic acid are frequently, and propionic acid, ascorbic acid or sorbic acid sometimes, added to processed fruit products, especially juices, to inhibit the growth of yeasts and molds.

b. Chemicals that delay ripening or senescence. Examples are: the kinins and kinetins that delay chlorophyll degradation and senescence in leafy vegetables, gibberellins that retard the ripening of tomatoes and hold citrus fruits on the tree beyond normal maturity, and auxins that delay physiochemical deterioration of oranges and green beans.

c. Growth retardants that inhibit sprouting and growth. Examples are maleic hydrazide which is applied pre-harvest and inhibits sprouting in a number of stored commodities, e.g., onions and potatoes. A number of chemicals are applied post-harvest to potatoes to control sprouting, for example, CIPC, TCNB and MENA. Daminozide (Alar) give a increased fruit firmness, bettor colour and early maturation in apples.

d. Chemicals that hasten ripening and senescence. Examples are ethylene and compounds such as Ephephon that release ethylene, abscisin, ascorbic acid, ß hydroxyethyl hydrazine (BOH), acetylene and substances that release acetylene such as calcium carbide, and certain alcohols and fatty acids.

e. Chemicals that may hasten or delay ripening and senescence depending on the dose and the commodity on which they are used. Examples are 2, 4-D; 2,4, 5-T; indoleacetic acid (IAA) and naphthalene acetic acid (NAA).

f. Metabolic inhibitors that block certain biochemical reactions that normally occur. Examples are cycloheximide, actinomycin D, vitamin K, maleic acid, ethylene oxide, and carbon monoxide.

g. Ethylene absorbants. These delay ripening and senescence because they remove the ethylene produced by the fruit. They are usually placed in clove proximity to the commodity and leave no residue on it. An example is potassium permanganate- impregnated alumina or vermiculite (fur fir).

h. Fumigants to control insects or sometimes molds. Ethylene dibromide and methyl bromide are the most commonly used fumigants.

i. Colouring. The use of artificial colours is sometimes permitted in order to improve the appearance of a fruit. For example, fresh grange a from Florida may have artificial colour added to the akin for cosmetic purposes. Since most people do not eat orange skins other than for marmalade it is considered to be a harmless addition.

j. In warm climates ethylene is used to degreen lemons, oranges and tangerines imparting a brighter colour to the skin. Ethylene is a naturally occurring metabolite of ripening fruits.

k. Food additives. A number of compounds are permitted to be added to processed horticultural products for preservative or functional effect. The major preservatives are sulphur dioxide, benzoic acid or benzoates, and sorbic acid or sorbates. Functional additives include antioxidants, colouring, flavouring, thickeners, emulsifiers, etc.

There are two important differences between the use of chemicals in the field and the use of post-harvest chemicals:

i. Smaller quantities of post-harvest chemicals are used. For example, the normal dose of CIPC for controlling sprouting of potatoes is about 30 grams per ton and the normal dose of ethylene dibromide for fumigation of fruit a and vegetables is about 30 grams per ton. These levels contrast with the use of field chemicals where doses of one to several kg. per hectare are commonly used.

ii. The chemicals are not broadcast over the field but are applied in the confined apace of the storage chamber.

It is impossible to obtain figures for the quantities of post-harvest chemicals that are used because this is considered proprietary information by the companies that manufacture and formulate them. However, all post-harvest chemicals are classed as "minor use" by the U.S. Environmental Protection Agency because the quantities used are relatively small.

In the U.S. a company must produce experimental evidence of the toxicity, safety, and usefulness of a new agricultural chemicals before it can be registered for use as an agricultural chemical. Each use must be cleared through registration for every commodity to which it is applied. The Environmental Protection Agency has the responsibility for registering pesticides and setting tolerances.

The chemical suppliers usually described in detail on the label and/or in supplementary literature exactly how, when, and how much of the chemical is to be used. This is backed up by the agricultural extension service of each state which keeps in close contact with the farmer. Most states have a cadre of inspectors who regularly draw samples of horticultural products and submit them to a central analytical laboratory to assay for chemical residues. Few violation of the regulations are detected, and most samples tested are found to be well below the tolerance. Most other developed countries maintain close supervision and control of the use of post-harvest chemicals on horticultural products.

The situation may be quite different in the less developed countries where governments usually do not have the expertise or back-up analytical laboratories to monitor adequately the use of post-harvest chemicals on perishable crops. It is difficult to obtain information on this topic, but from a general knowledge of how governments in LDCs operate, it appears to be a matter that deserves investigation. Presumably, the pesticide tolerances for the major export crops (e.g., fresh bananas) are effectively monitored by the large corporations who operate this trade and by the developed countries that import these commodities. One cannot be so sanguine about the horticultural crops that are indigenously produced and consumed in the less developed countries.

The whole issue of harmful chemicals in the environment is complex and not very clear at the present time. However, we can make two reasonable assupmtions about chemicals added to horticultural products:

1. The developed countries have the expertise to engage in the debate on the harmfulness of chemicals, to evaluate the risks and benefits of their use, to enact new legislation controlling their use as new knowledge becomes available and to establish the inspection-analysis-prosecution machinery to ensure that the legislative intent is carried out. It is reasonable to assume that any proven grave rick from the use of chemicals will soon be brought under control in the developed countries.
2. Most of the less developed countries have little of the kind of expertise listed above. There is a real risk that chemicals will be improperly used and their improper use will not be brought under control, with the potential to cause harm to human health and the environment. Therefore, the use of post-harvest chemicals in a given country should be discouraged until adequate inspection services analytical laboratories have been established to ensure that these chemicals are used safely.

Environmental effects. Misuse of certain post-harvest chemicals may lead to seriuos environmental harm. As far as can be determined little is known about the ultimate effects of post-harvest chemicals on the environment when they are correctly used. It seems to be generally assumed that since the compounds are used in small quantities in confined areas, and since most of then decompose into nonactive substances there are no adverse environmental effects. Although this is a reasonable assumption there is little concrete evidence either for or against this widely held opinion.

Pests

The major causes of lose in perishable produce after harvest are certain pathogenic fungi and bacteria. Viruses and nematodes play a minor role in postharvest losses; rodents and insects are also generally of lesser importance in contrast to the significant damage they cause in food grains.

In addition to the direct lone in quantity of food resulting from microbial infections, a partial loss results because of effect on appearance and/or quality resulting from disfiguring surface infections of fruit, root, and tuber crops. Other secondary adverse effects may include a decline in shelf life, possible contamination with mycotoxins, acceleration of ripening because of release of ethylene in pathogenesis by certain fungi, and in some instances deterioration of canned fruit crops because of the presence of heat-resistant hydrolytic enzymes formed by decay fungi in fruit tissues prior to the canning process.

The loss in the post-harvest period may originate from infections that were initiated by fungi during the growing season well in advance of harvest. Much of this pre-harvest infection involves a group of fungi that are capable of infecting healthy developing fruits either by direct penetration, e.g., anthracnose deseases caused by species of colletctrichum or by invasion via natural openings much as lenticels or stomates or through breaks in the tissue at the points of attachment of fruits to the plant. In many cases the infection process may be incomplete. Thus, sub-cuticular mycelium may be formed which remains in a latent stage until the post-harvest period when changes in susceptibility may occur and the pathogen mycelium may ramify through the tissue.

Many of the fungi (e.g., species of Penicillium, Rhizopus, and Gectrichum) and bacteria (e.g., species of Erwinia, Bacillusand sometimes Clostridium) involved in decay problems associated with the post-harvest period may be considered as opportunistic pathogens. They are usually incapable of penetrating unijured tissue or aggressively attacking vigorous healthy plants during their active growth period. However, they do have the ability to parasities fleshy plant organs when tissues are bruised, injured by insects, or otherwise placed under environmental stress. In many cases tissues are invaded by a succession of organisms which may interact in a synergistic manner.

Each species of root, tuber or fruit is affected by specific groups of fungal or bacterial pathogens. It is important that broad non-specific designations of these organisms be avoided (i.e., molds or rot organisms). The species of fungi or bacteria associated with specific decay problems should be properly identified. For example, it has been clearly shown that species of Knisopus differ markedly in their susceptibility to specific fungicides.

The main factors affecting disease development are:

1. Most susceptibility
2. maturity
3. wounds and wound healing
4. temperature of the commodity
5. relative humidity (especially in storage)
6. packaging
7. handling in general
8. concentration of inoculum

The basic methods of control involve three different approaches 1) prevention of infection 2) elimination of incipient or latent infections, and 3) prevention of spread of the pathogen in the host tissue.

Losses due to micro-organisms may be reduced by:

1. refrigeration
2. improved handling procedures
3. pre- and post-harvest chemical control
4. Although every effort needs to be made to minimize or reduce dependence on chemicals, in many cases no viable alternatives exist. A wide range of fungicides are now available that are effective and safe to use. Incipient infections can be eliminated or reduced by application of certain fungicides such as benomyl which have the capacity of diffusing into host tissue and killing the pathogen in situ. Certain fungal pathogens initiate infections of fruits and vegetables during the growing season and these can be controlled best by timely application of fungicides prior to harvest. It is essential to ensure that chemical controls are used properly.
5. preventing contamination during the washing process
6. In the case of tomatoes and possibly other fleshy vegetables when warm product is washed in cold water infection is augmented. Because of the temperature differential air in the tissue contracts and this draws water, often containing soft rot bacteria' into the tomato via wound e

and fresh stem scars. If soft rot bacteria are present, water deeper than about 0.3 m increases the risk of infection because of the pressure differentials inside and outside the fruit. If washing is needed' proper, prompt drying is essential in order to prevent rapid growth of spoilage organisms in superficial wounds and lenticels. Sweet potatoes and yams have a longer shelf life when stored unwashed.

Unresolved problems

The actual physiological processes involved in the rapid deterioration of cassava after harvest are act yet fully understood, although much research is in hand. For many other commodities also, knowledge is lacking about changes in physiological processes in the post-harvest period. In particular there is a need to determine the relationships of those changes to the increased susceptibility of perishable products in the pool-harvest period.

Date is lacking with respect to the effect of chemicals, singly or in combination, AS used in the growth period on the storability of commodities. Fertilizer, weed control chemicals and vine killers for potatoes are examples of the chemicals used which may affect storage characteristics or disease susceptibility

The physiological basis for resistance of perishables to post attack needs to be known. Mechanisms of resistance of micro-organisms to different fungicides and bactericides are also little understood.

Biological control systems should be further explored. For example, it is known that, by dipping root stocks into suspensions of an avirulent strain of the grown gall bacterium that produces a very specific antibiotic (bacteriocin) later infection by the grown gall (bacterium) of stone fruits can be prevented. Other similar antagonistic relationships are known to exist: they may offer possibilities for more techniques for biological control of insects. Biological controls are an important feature in the plant protection programmes of the People's Republic of China.

Marketing and distribution

The approach of the market specialist to the problem of food losses in perishables is to identify the place in the market chain where losses of unusual magnitude occur. The physical place where such losses are registered is less important than the position in the chain whore losses occur and the relation of the specific loss situation to the total market chain. In relating losses incurred at a specific situation to the total chain, the objective is to find out whether the lose can be explained by defects further up or down the chain or by the system as a whole.

Losses in perishable produce occur everywhere from the field to the ultimate consumer and depend on the degree of perishability of the produce; they are inherent in the very nature of the product. Since the market chain or system refers to specific operations, handling, transportation and trade practices, there is a close correlation between the type and magnitude of loss incurred by a specific product and the chain or system wherein it moves. This implies that for a given commodity moving in a particular chain, there is something like a standard loyal of loss inherent in the chain, the reduction of which could not be obtained by improving isolated operations taking place within the chain. It would require a change in the total market system itself. To avoid losses or even significantly reduce them et isolated stations in the chain may not be a realistic proposition. However, any obvious isolated practice that leads to heavy losses, such as faulty packaging, must be corrected.

Marketing methods and conditions vary widely from country to country and any attempt to attribute losses to a particular point in marketing chains or to any specific system or marketing runs into difficulty because it is not possible to generalize on a wide basis. For this reason, a systems approach should be adopted for dealing with food losses whereby all the factors applicable to a given situation and in an individual country have to be considered before any meaningful diagnoses can be made. Any success achieved in reducing losses at the "grass roots" level particularly where applied to traditional marketing systems should receive wide publicity as such successes are not often known beyond their immediate area of application.

Management of losses is essentially action-oriented. It is effected within a given market system for which "norms" of foes lovers can be established. Managing the losses with reference to acceptable levels would be quite similar to management by objectives. Each marketing system has its own rationale and is affected by policy decisions with regard to production, marketing and consumption. The latter will have a direct incidence on the effect of any measures that might be implemented for the reduction of losses. An examination of the type of production and marketing system would be helpful in revealing how susceptible a given market system or sub-system may be for the introduction and application of measures aimed at bringing losses down to a desirable level.

Traditional subsistence systems are widespread throughout the developing world sad are characterized by local exchange or barter trade, sharply limited geographical movements of produce and typical small units of sale or barter. Losses do occur but they tend to be overestimated and may be difficult to reduce. The subsistence economy sets its own limits: the chain from field to consumer is usually short, both in time and distance. Practically everything is consumed because every quality finds a ready consumer. Improvement of the system in terms of reduction of losses may be limited to the provision of shade. Recommendations might not go beyond encouraging the producer to collect his fruits and vegetables under a tree or build a make-shift shed or tent.

The situation is quite different when one looks at the emerging marketing systems where a surplus production economy is rapidly replacing the previous subsistence or semi-subsistence economy and the marketing system is not well adapted for the teak of collecting, moving and distributing massive surplus production. Losses of unusual magnitude do regularly occur in the system and here there is good scope for the introduction and application of specific measures to reduce losses depending on the readiness of the systems to transfer the improvement down or up the chain. Most measures that focus on gentler handling, better conditioning, faster transportation and proper storage would seem to be effectively applicable only within an improved market infrastructure including suitable roads. Quality consciousness and the introduction and acceptance of some forms of quality separation by the trade and the consumer must precede the demand by the farmer for bottle, and therefore higher cost, boxes or containers. This again can hardly be expected if the higher prices paid for better quality cannot be transferred down the chain to the producer. Much closer communication relating producer capacity to retailer/consumer demand is a prerequisite to such developments, sad the more efficient management of marketing enterprises.

In some regions particularly high levels of lose have been observed when governmental and quasi governmental agencies participate directly in horticultural crop marketing. This to a great extent is due to faulty management inexperience in the commissioning and operation of cold stores, lack of technical knowledge in despatch and transportation techniques, the absence of a sales operation to set oh the buying programme and generally unrealistic pricing policy. The scope for remedial action in these incompletely integrated sub-marketing systems is great. This may deserve priority in any action programme for the reduction of losses. The crops which have typically been the object of intervention programmes by Governments are potatoes and onions, but other crops such as oranges have in vein. areas been included.

Socio-economic aspects

A complete analysis of the post-harvest system would include not only the "physical" but the "human" aspects; that is, a study of those social, psychological and economic factors which influence the behaviour of the people involved in post-harvest activities. It is these people, the subsistance farming families or those individuals who produce, process or market on a commercial basis, who are required to change their behaviour when schemes are launched to reduce post-harvest losses. Any technological change required of a producer is likely to produce social and/or economic changes. Additionally, the nutritional status, e.g., of children or mothers, may suffer, Introduction of even simple machinery can displace women's paid labour; production of a marketable surplus may provide the family's first cash income. Preliminary diagnostic studies make it possible to ascertain the possible effects of such technological changes and to identify the socio-economic constraints influencing the producers and their acceptance of proposed changes. The prime task of such studies would be to locate where losses occur, who is responsible and who bears the loss. A range of social and economic relationships also needs to be explored with respect to any proposed remedies.

Despite the fact that the producers' attitudes to lose and understanding of its implications may be limiting factors in motivating the adoption of change, their skills and knowledge ought not to be ignored or underrated. All concerned with post-harvest operations should be involved in the planning of improvements.

It is of particular importance to identify which members of the producer's, handler's or processor's family are responsible for the various post-harvest activities. Efforts can then be made to ensure that the right people are included in training programmes, credit schemes and other services which may be planned.

Is order to make a systematic approach to the reduction of post-harvest losses it may be seen that co-ordination is necessary with activities developed to alleviate related problems . A post-harvest system cannot be vi owed in isolation from ether aspects of rural development. In this connection the Programme of Action adopted at the 1979 World Conference on Agrarian Reform and Rural Development provides guidance regarding priorities for action.

FUTURE DEVELOPMENTS FOR HORTICULTURAL PRODUCTS

Brand new concepts that eventually lead to extensive new technologies usually burst upon us without warning. It is impossible to forecast when, or in what form such major breakthroughs will appear.

Processing

Rising standards of living can be expected to increase demand for canned and frozen horticultural crops as well as the demand for convenience foods that contain one or more horticultural crops as a major component. Losses after processing are usually very low. On the other hand, they are energy-intensive and the disposal of the discarded packaging materials as solid waste creates some environmental problems.

Large scale factory processing of horticultural crops concentrates large quantities of waste materials much as skins, cores, and seeds in one place in contrast to the highly dispersed distribution of waste from commodities that are consumed fresh at home. Traditionally, little attention has been given to dispose. of this waste material and it has caused environmental problems. In recent years, attention has been drawn to this problem, e.g., the joint UNEP/FAO sponsored seminar on agricultural residues held in Rome in January 1977. Many governments now require food processing plants to upgrade the handling of their waste materials.

New Chemicals

The large chemical corporations continue to search for new chemicals to control the pests and diseases that attach crops and products. The cost of testing mud. registering a new agricultural chemical is now SO high that only the larger corporations can afford to take the risk, and even they are mainly interested in pesticides for major crops such as maize, wheat, soybeans and cotton because they cannot expect to recover the costs of developing a new chemical unless potential sales are high. After a chemical has been registered for use with a major crop, it can also be registered for a minor crop at much leas cost if it is shown to be effective.

New chemicals for post-harvest treatment of horticultural crops will continue to appear but mainly as spin-offs from use on major crops. It is unlikely that many new chemicals will be developed especially for post-harvest application because of the relatively low volume of sales anticipated for such application, which is insufficient to justify the high cost of obtaining the initial registration.

Irradiation

Irradiation of horticultural products kills infesting insects thus permitting products to be shipped into areas that have a quarantine against certain insects. It also delays sprouting of bulbs and tubers, permitting long-term storage of commodities such as onions, potatoes and yams without sprouting.

Although several decades of research have been devoted to this peaceful use of atomic energy there is almost no commercial use of irradiation technology even though it continues a to be energetically advocated by the International Atomic Energy Commission and a total Or 26 commodities have received restricted or unrestricted clearance in on or more of 19 countries having legislation on irradiated food. Maxie et al concluded after a lengthy study that irradiation was, in general, not as effective as good commercial refrigerated storage for fruits and vegetables. A 1976 report point d out that after a quarter of a century Of world-wide work on this promising technology there is today only one truly commercial application, namely, the inhibition of potato sprouting in Japan.

Radiation technology has not been wider, adapted in commercial practice for the following reasons:

a. It costs more than chemical treatments. Chemical treatments usually cost about 10 cents to $1.00 par ton of product. Irradiation probably, costs several times as much although contact figures are difficult to obtain. The twelfth session of the Joint FAO/WHO Codex Alimentarius Commission in 1978 noted that the economic feasibility of the irradiation process still required practical demonstration.

b. The rood must be passed through the irradiator thus creating an additional handling stage which adds to the cost, and may increase bruising and wounding of the commodity.

c. Irradiated vegetables may be more susceptible to storage rots and fungi because the natural would bearing processes are impaired. Irradiated potatoes sometimes darken when cooked. Irradiated onions contain a small piece of unsightly black tissue in the centre of the onion a bare the growing tip is killed by the radiation.

d. Much more elaborate safety precautions are required for an irradiator than for chemical treatments.

e. The softening of vegetable tissue caused by irradiation makes the commodity more liable to bruising during subsequent handling and transportation.

To sum up, it appears that irradiation technology for horticultural products is unlikely to be widely used under present circumstances. Although the International Atomic Energy Commission continues to maintain a vigorous programme to introduce irradiation as a food preservation technique it seems unlikely that irradiation will be widely used unless all chemical treatments are banned or some now major breakthrough in irradiation technology is developed.

Drying Technology

Since drying is the lowest cost preservation technology it should develop as a major method of preservation of horticultural products especially in LDCs. Much drying can be accomplished by means of solar energy. Simple drying systems can be established as small-scale localized unite for single families or villages. The fact that the quality of dried products is not necessarily as high as the canned or frozen product need not be a liability in developing countries where low price is so important and the expectation of quality and convenience is not as high as in the developed countries.

ROOTS AND TUBER CROPS

The root/tuber crop resource

Useful insights into the post-harvest loss complex of root crops are provided by a study of their characteristics. The principal roots and tubers are derived from nine species. They are all of high Moisture content, and their parenchyma cells are packed with starch grains. These materials show variable degrees of inherent keeping life, from some species of yam, like

Dioscorea alata which remains dormant for 3-4 months to cassava in which there is no natural dormancy. When such material is stored in the fresh state, respiration and transpiration continue with inevitable losses of water and dry matter. Prolonged high levels of transpiration result in a change of texture affecting quality and weight loss. The dry matter loss, caused by superation constitutes a real food value as distinct from the moisture loss which only has an economic value. Sprouting at the end of dormancy can result in dramatic loss as the physiological state is altered. At this stage the stored starch is transformed to sugars and utilized by the elongating shoots with appreciable loss of both food value and moisture.

The main root/tuber crops are cassava (Manihot esculenta Crantz), the yam (Dioscorea spp.), the potato (Solanum spp.), the sweet potato (Ipomoea batatas L. (Lam) and the edible aroids (Colocasia spp. and Xanthosama sagittifolium). With the yam, the potato, and the aroids many different species are used in the cultivated complex. Differences between these are usually sufficient to merit separate treatment.

Tare (Colocasia esculenta var esculenta) illustrates the system of maintenance of continuous food supply well. This species which has no seasonal growth constraint is planted in moist situations conducive to year-round growth. By systematic replanting whenever material is harvested, continuous supplies are assured. A similar absence of seasonality in cassava (Manihot esculenta) together with its ability to produce mature roots at various time intervals after planting permits year-round harvesting in most ecological systems. By opting for continuous food supply the opportunity for post-production losses is restricted to the food preparation phase. The same benefits accrue from the use of crop scheduling

Maintenance of food supplies ever restricted periods through storage is an essential feature of yam-based food supply systems. The success with yam storage lies in recognition of the limits to the potential for storage and in producing an environment conducive to keeping the material in sound condition. Perceptive farmers like those in Nigeria and the Ivory Coast often construct different stores to meet the separate requirements of D. alata and the different culivars in the D.rotundata/cavenesie complex. These systems work well within the normal period of yam tuber dormancy. They are so successful in West Africa that improvements in the beat indigenous practice could hardly be expected without recourse to the technologies of reduced temperature storage, the use of controlled gas environments or ionising radiation. These latter technologies are however not commonly applicable in developing countries because of cost and lack of know-how.

Considerable quantities of roots and tubers are transformed into more durable products by drying, fermentation and commination in different combinations and sequences producing a variety of materials each with distinctive characteristics. Other benefits that often accrue from transformation include the removal of toxic substances naturally present and the convenient-to-use nature of the transformed product. Although the ultimate keeping potential is constrained by environmental conditions, transformations have good prospects for long-form maintenance of root/tuber crop supply because they convert a perishable commodity into a form with stability similar to that of the cereal grains.

A feature of much of the indigenous processing technology is its high labour requirement. Because of this, food supply strategies based on transformation tend to be restricted to economic environments where labour is available. As a result root crop transformation is widely practiced among subsistence economies whore labour inputs have zero or low cash value. Attempts are being made to establish root/tuber crop transformation commercially, but success has been variable. Radical changes in the entire system of production may be required, as has been suggested with yam (D. rotundata/cavonenais) in Nigeria for manufacture of instant pounded yam.

The major root/tuber crops

Cassava (Manihot esculenta Crantz)

Known variously as manioc, yuca, tapioca and mandioca, the edible portion of cassava is a starchy root which matures in harvestable state in 8-24 months according to cultivar and climate. In addition to the roots, the leaves also have potential for use as food as for instance in pert e of Central and West Africa. Cassava cultivars are often grouped into "sweet" with relatively low contorts of cyanogenic glucosides and "bitter" with high cyanogenic glucoside content although many intermediate forms exist. The "sweet" types may be eaten raw or lightly boiled without harm while the "bitter" forma require processing to remove cyanogenic glucosides.

The cassava roots have no natural dormancy and are highly susceptible to deterioration. The nature of this deterioration is as yet not fully understood, although two factors, one an indigenous physiological one and the other associated with microbiological infection, have been identified. The physiological deterioration' of which the symptoms are tissue discolouration, can commence as soon as 24 hours after harvest. Acting independently of, but usually following, the physiological changes, microorganisms, both fungi and bacteria develop in the flesh of the roots causing additional damage and loss: this microbial deterioration usually develops only 5-7 days after harvest.

This rapid post-harvest deterioration of cassava roots places serious constraints on their use with fresh produce trade and on the holding of buffer stocks for large-scale processing.

The short storage life of fresh cassava imposes constraints on its distribution and use. As a result, the choice of market for which production is intended becomes influenced by location. The range over which production for the fresh market extends is therefore a function of the distance from their markets and the efficiency of transport. The time of delivery rather than distance per se is the important limiting factor. In the vicinity of Bogota, Colombia, up to an 8-hour interval from harvest to delivery is acceptable. With the current status of road a and vehicles this permits production within a 300 km radius. With more effective preservation or a faster delivery system, this range could be extended. It is claimed that the market for fresh cassava in Colombia is so structured that as quality deteriorates, the price is lowered, thereby enabling roots of decreasing quality to be bought by successively poorer income groups.

Techniques to extend the fresh life of cassava roots have been independently developed in several countries. At the Instituto de Investigaciones Tecnologicas in Colombia deterioration has been delayed by coating tubers with a film of paraffin wax (IIT 1973). At CIAT fresh tubers were shown to have improved keeping characteristics when plant stoma are out and removed but tubers are loft unharvested for up to 14 days. Tubers so treated remained sound for as long as 30 days. This technique has much to commend it, but has not yet been tested on a large scale. Other practices developed in joint CIAT/TPI projects include storage in earth silos' and storage of roots in plastic bags and in absorbent packs. These methods appear to have some utility in small and medium scale production but have not yet been proved in large-scale commercial operations. None of those techniques have as yet attained general application.

The transformation of cassava to more durable forms, frequently in forms more convenient to use than fresh roots, is partly associated with the detoxification process for the removal of hydrocyanic acid. From the crop loss point of view these products are important since they are frequently more able to avoid the rapid deterioration of the fresh tubers. Cassava products may be classified in a variety of ways, but for simplicity these may be regarded as food starch ant whole root products. These latter may be fermented or unfermented and may either be "flour" or "gari" types.

Cassava starch is extracted from roots after pealing, washing, reaping, squeezing settling and drying. Two types of starch are produced, sour or fermented and sweet or unfermented. Sweet starch is dried immediately following extraction while sour stanch is stored for 3-4 weeks when some fermentation involving lactobacilli and yeasts occurs. Sour starch is favoured for baking and commands a better price than sweet starch at least in some countries. This latter has similar conformation to corn starch with which it competes disadvantageously.

Gari, the most commonly used form of cassava in West Africa, accounts for some 70% of the entire cassava production in Nigeria. It is estimated that between 4 and 5 million tone of roots are used each year for this purpose. Although there are close parallels between the production of gari and of Brazilian "farinha de mandioca", the West African variant has some distinctive features.

The effect of the longer fermentation affects the taste by the production of lactic acid, as wall as well as reducing the content of free hydrocyanic acid. Production at levels of indigenous technology is highly labour-intensive, but this constraint is being progressively removed through a variety of innovations aimed at increasing labour productivity. The hygroscopic nature of gari is a major constraint to its keeping quality. 1a a humid atmosphere it can absorb sufficient moisture to make it vulnerable to the growth of fungal organisms. The recent practice of packaging gari dried to a safe moisture contort of around 12% in sealed polyethylene bags enables the product to be kept in good condition for extended periods.

Farinha de mandioca is the Brazilian product from which gari derives. The stages of preparation are similar to those for gari, the roots being washed, peeled, grated, pressed, toasted and classified by size. Unlike gari, the mashed or grated mass is not permitted to ferment. The heating dries the product and allows some gelatinization and dextrinization of the starch as also occurs in gari making.

Tam (Dioscorea Spp.)

With yams, storage life is determined by dormancy since attempts to store yams after sprouting has commenced are impractical. During the period of dormancy the metabolic processes continue at reduced rate thereby keeping dry netter losses at relatively low levels. The various yam species have different periods of dormancy with 3-6 months for D. rotundataand D. alata and 1-2 months with D. cavenensis or D. trifida. Pathological factors are also of great importance in causing postharvest losses in yams.

Traditionally, in West Africa, yams are stored in "yam barns" where good but uncontrolled conditions for storage exist by building the barns in the deepest available shade. This prevents heating by the nun and also helps to maintain high humidity. Several other indigenous yam storage systems are used but moat of these are less satisfactory than the "barn" or "claie verticale".

It is generally believed that these systems are well adept d to the yam varieties and to the climatic conditions prevailing in West Africa and offer little scope for improvement at current levels of technology.

Of the methods used to extend storage only the use of ionising radiation has been an unreserved technical success. The practical application of this technique is not yet generally available. Reduced temperature storage is limited by possibility of chilling damage around 10-12 C and by the growth of fungal organisms at temperatures around 14° C. Treatment with chemicals used to suppress sprouting in other crops has had little success but active research is being pursued by TPI. Controlled gas storage has not thoroughly been investigated.

Transformation of yams to more durable forma is possible but these methods have not yet met with much commercial success. Traditional products such as "amala" are variable. This may be related to the intrusion of undesirable dark colours into the final product from the use of a mixture of varieties and the failure to deactivate the polyphenol oxidases during processing rather than to any inherent basic defect of the process. The attempts of commercial organizations to produce an instant pounded yam failed more as a result of nontechnical problem such as the logistics of raw material supply than to technical defects in the process. New processes recently developed at the University of Ife, Nigeria, may provide the required answer. A yam flake product developed to pilot project level in the West Indies has not continued long in production at a commercial level.

The Potato (Solanum Spp.)

The potato, Solanum tuberosum, originally of highland tropical origin, but developed mainly as a temperate crop, is now being developed as a lowland tropical crop. Selections tolerant to high temperature and short days are being made in several parts of the tropics. At higher elevations in the tropics it is already established as a crop of some importance.

She extension of potato growing to these areas is a major concern of International Potato Centre. This Centre works together with existing bodies selecting areas for activity on the basis of comparative advantages. This approach forms a rational basis for co-operation and avoids the sometimes unintended, but nonetheless real, tendancy towards replacement of national effort with that of an international organization.

The work on post-harvest lose in this Centre is integrated with that of the social science unit permitting a broader view of the problem. This approach, taking account of social, economic and ecological factors, could provide useful guidelines in the development of other work on post-production losses. An account of the studies in the Mantaro Valley is given by way of an illustration of the CIP approach.

The Mantaro Valley lying some 300 km to the South-East of Lima has been described by Mayer. It is a riverine valley acme 60 km in length and from 2 to 22 ho in breath. Three broad ecological zones are recognized. The valley floor is some 3,200 - 3,450 m above sea level. The eastern slopes between 3,400 and 3,950 m and the western slopes of similar elevation.

Finally a high zone above 3,900 m. The flat riverine soils of the valley floor have a high productivity with a capacity to support a diversified agriculture and a high population density. In this zone a contrast between large and small farm a is noticed. In the zone of sloping land above 3,450 m soil productivity declines as dose the number of crop species. The practices on the East and West slopes differ. On eastern slopes potatoes maintain their place in cropping systems but play a diminished role on western slopes. Under the more severe conditions of the High Zone potatoes become the most important component of the cropping system occupying some 57% of the arable land. These zones are constrained by three main environmental factors: precipitation, temperature and water balance. These are all subject to changes relative to elevation. Generally precipitation and exposure to frost increases with increasing elevation.

Several products are made from potatoes including chuno, papa seca, tungush and potato starch. Ecological zone and economic circumstances result in different comparative advantages between choice of product. The most common product at low elevation is papa seca, a parboiled dried product which requires, fuel for boiling but which utilized what would otherwise be a waste product. This restricts production to the lower zones where fuel supplies are more assured. Hybrid or native varieties are used for papa seca since this process would not be sufficient to remove the glycoalkaloids present in the bitter varieties. Processing into chuno or tongosh which eliminates the bitter principle requires no cooking. The process however requires heavy night frosts and open areas on which to expose the potatoes to natural freeze drying. These conditions are all readily met in the high blue. Locally produced potato starch competes with commercially produced corn starch. It is clear from this example that changes due to the technological superiority of one process over another is unlikely since a different technique is adapted to meet different complexes of factors.

Potato Storage

Potatoes are stored for use as seed or for food or sale. Different methods of storage have been described by CIP from which the following summary has been taken. Ho recognizes three main systems: house' out-building and field. Peruvian rural houses according to Stein appear to have been cosigned with storage of agricultural produce in view. The convenience and security of storage in the house is evident. The actual technique of storage adopted appears related to the itended use for the potatoes. Potatoes to be boiled in their skins are kept is the attic where the conditions permit drying and some shrivelling, causing the potatoes to become sugary and thus more favoured. Potatoes which are to be pealed before cooking may be stored on the ground where they love leas moisture and HO remain firm.

Seed potatoes may be left on the ground on a thin layer of straw or eucalyptus leaves, in piles against the walls of houses or in shallow bins or "trojas". Where animals may have access to them if they are loft on the ground, the potatoes may be placed on a platform or chaclanka of eucalyptus branches in the rafters and covered lightly with et raw.

Field storage is more frequently adopted at higher evaluations. Here strawlined clamps, covered in turn with straw and soil, may be used. The cool moist conditions reduce dehydration but the system is vulnerable to flooding which may result in dramatic crop losses. At elevations of about 3,000 m ventilation with air at ambient temperature assists the storage process.

An interesting series of low-cost potato stores have been developed by the storage unit at CIP. These are developed and adapted to take advantage of prevailing conditions to facilitate storing. Control over the storage environment is attempted by using natural ventilation systems coupled with adequate insulation and shading. Where appropriate humidity modification using simple measures is also being researched. In the case of seed tubers, these are stored in simple low-cost stores exposed to natural diffused light which, to some extent, replaces the need for controlled low temperature storage.

Sweet Potato (Ipomoea batatas (L.)

Sweet potato is a crop of considerable unrealized potential. The roots are widely used as a carbohydrate food and recently attention has also been drawn to the nutritional value of its leaves as food. The ability of the roots to become suberized at high temperature and humidity in the classical curing process, gives the crop potential for extended storage. Storage is particularly successful when kept at relatively cool temperatures, although, it held below 12°C, chilling damage can occur. The most successful storage occurs in the southern states of America and in Japan, and successful storage for extended periods has still to be achieved in the Tropics.

There is interest in transforming the roots to more durable forms. In Trinidad, at the University of the West Indies, a sweet-potato-based flour was shown to have utility in wheat flour substitution in bakery products. There is interest there and elsewhere in the preparation of breakfast foods from sweet potato tubers. The manufacture of dried sweet potato flakes is already an established industry in the United States.

Taro (Colocasia esculenta) and Other Edible Avoids

The edible corms of these crops are normally eaten fresh after boiling or baking. Estimates in Fiji showed high peeling and trimming losses during preparation as food. The fresh corms to not normally remain sound for long periods, and signs of deterioration become apparent a few days after harvest. Observations in Fiji indicated that packaging in plastic bags enhances keeping quality. In the South Pacific it is traditional to keep taro by transforming it into a fermented product of which "Poi" is the Hawaiian variant. It is of interest to note that the presence of oxalic acid raphides in the corms acts as a deterrent to rats.

Fruits and Vegetables

General Considerations

Many post-harvest losses are direct results of factors before harvest. Fruit a and vegetables that are infected with pests and diseases, inappropriately irrigated and fertilized, or generally of poor quality before harvesting, can never be improved by post-harvest treatments. Very often the rate of commodity loss is faster if the quality at harvest is below standard. Thus, the processes in the attainment and maintenance of quality from production, harvesting, handling and marketing must be considered a unified system. The success of preserving the harvest-fresh quality of produce demands control of each step in the system, depends on the previous stop and therefore is a chain of interdependent activities.

The small size and isolation of many vegetable farms make it logical to employ manual harvest labour which is often relatively cheap and non-organized. Traditional methods of harvesting are still employed using a minimum of mechanical aide. Many of the farmers may know their market but do not analyze it and plant without much concern for it. Price considerations are usually given more weight than the quality of the produce. Early harvesting of carrots, chayote, snap beans, squash and bottle gourd give better quality, but lower yield. On the other hand, vegetables may be harvested before reaching prime quality if the prevailing market price is high duo to the scarcity of the product. Knowledge of maturity indices is often inadequate in moat instances visual indices being used. Therefore, more experienced farmers can deliver better quality products than those with less experience.

The deterioration of a product starts during the harvesting operation. The more carefully a product is handled, the slower the deterioration process during subsequent handling operations. However, the farmers mar be unaware or indifferent to the condition of the product after harvest, and harvesting procedures may thus be rather careless. The only constraint is to avoid external injury. Until farmers are convinced that careful handling will increase profits, it will be difficult to persuade them otherwise.

In many countries contract buying is practiced where the contractor takes charge of the harvesting and may exert strict supervision of the operation.

In meet of the developing countries the world, harvesting methods are very simple. Picking poles, to which a hook or cutting knife is attached' are generally used. Fruit blemishes and injury are usually unavoidable. Quite often the fruits fall to the ground becoming subject to field infestation. Planning, preparation and organization in the harvesting operations are necessary. Suitable harvesting tools, hand gloves, containers and supplies are needed by the harvesters. The grower should exert strict supervision on the harvesting operation.

Grading is a thing that most farmers are loath to do until they are convinced it will bring them added revenues or increases the acceptance of their products. A farmer may separate different varieties if they are distinguishable, but consumers in the tropical countries usually are more price conscious than quality conscious. That is not to say that they are unmindful of the quality, but if good quality products are too highly priced, they are often willing to settle for a poorer quality product. With the variety of vegetables offered for sale, shore is usually no difficulty in Betting cheaper substitutes. Thus, initially, for local markets, grading should be limited to only what is necessary. Deformed fruits and those with splits, punctures and incipient rotting should be removed. If the market is conditioned to accepting sound fruit with harmless surface blemishes (e.g., russet, windscarring, or mite injury), it is well not to attempt to change the situation for the time being. The following are likely to occur in the absence of proper sorting and grading procedures:

i. Presence of rotten items which contaminate food products at later handling stages.

ii. Customer deception wherein good produce on top of the containers conceals items of low quality at the bottom of the pack.

iii. Growers do not maximize their income because best prices for their product are not obtained.

Containerization is probably, the weakest aspect in the distribution chain of fruits and vegetables. In general, packaging materials of unsuitable quality, such as large sacks, rough wooden boxes, second-hand cartons, bamboo baskets, or rattan containers are liable to cause produce to suffer from bruising, crushing' and puncturing. The use of suitable containers alone will do much to maintain the quality of fruits and vegetables. Farmers often selected the cheapest and most readily available containers. For example, the use of bamboo baskets has several disadvantages. The sides are sharp and easily bruise the produce. They are too deep and without sufficient side reinforcement, thus causing the produce to be Jarred or compressed. Moreover, handlers tend to throw rather than lift the produce gently because of its weight. The use of wooden crates could solve many of the objections to the bamboo baskets. The main objection to a wooden crate.

To summarize, transport losses are due to the following:

a. Unsuitable transport containers;
b. Overloading of mixed fruits and vegetables (in some developing countries people and even animals ride on top of the load);
c. Irresponsible driving;
d. Lack of feeder roads loading to highways or collection centres;
e. Rough roads;
f. Heat accumulation or very poor ventilation within the transport vehicles;
g. Virtual absence of refrigerated and insulated trucks;
h. Delays in product procurement after harvesting or at collection centres.

In many developing countries wholesale markets, if any are integrated with retail markets. Ideally, wholesale markets should be separated from the retail ones. However, this is influenced by the structure of the marketing chain and the scale of the industry. Frequently wholesale markets are overcrowded, unsanitary and lack suitable facilities for display, storage, ripening, loading and unloading. A main contributory factor to loss of leafy vegetables is trimming at wholesale markets prior to delivery at retail stands. Examination of trimmings showed that wholesalers and retailers trimmed their produce mainly because of the presence of decaying portions of leaves, due to bacterial soft rot. Thus, prevention of rotting during transport and storage would result in substantial reduction for postharvest losses of leafy vegetables.

Storage of fresh fruits and vegetables prolongs their usefulness, checks market gluts, provides a wider selection of fruits and vegetables throughout the year, helps in orderly marketing and may increase the financial gain to the producer. Adequate storage may reduce subsequent losses, but cannot overcome pre-storage losses. Adequate storage involves proper regulation of temperature, humidity, air circulation, proper stacking pattern, regular inspection, and prompt produce disposal as soon as maximum storage life has been attained. The feasibility of the construction of cold storage facilities and the interest among farmers and handlers to utilize the cold stores would again depend upon the economics parameters of the project. One of the greatest impediments to preserving quality through refrigerated storage is the consumers' strong preference for freshly harvested produce and resistance to stored produce.

Information on the storage temperature and humidity requirements of fruits and vegetables and the length of time they can be kept without decline in market value is either inadequate or unknown to those who need the information. If a farmer is persuaded to store his produce in cold storage and the market value decreases due to inadequate knowledge of the proper utilization of cold storage, it is not only he who will become disillusioned, but his friends also will be convinced of the non-profitability of cold storage.

Lack of capital may also force farmers to ignore the use of cold storage, even when available and effectively managed. Many growers depend on almost daily sales for their income and hence may be forced to accept a lower price immediately, rather than to store their produce in the anticipation of a higher price. There is also the storage rental price which the farmer may not be willing to pay unless he is thoroughly convinced that he will not only recover his investment, but will also profit.

The retailer usually disposes of produce which has been damaged by factors occurring further back in the marketing chain. At this stage, deterioration of perishables has already progressed to such an extent that street vendors have little opportunity to prevent further losses, either through storage or other preventive measures, except by estimating their potential sale for the day and buying only the amount that can be disposed of.

In Malaysia, although fruits and vegetable are sold in the open market, retailers improvise beach umbrellas arranged side by aide to provide acceptable shade to the commodity during market days. After the market is over in the afternoon the umbrellas are folded and the market area is again an open space. This simple technique or similar ones could be adopted in other developing countries as a temporary measure to provide shade to the produce.

Miscellaneous losses are numerous. The moat important ones are:

a. Over-purchase of cheap but highly perishable fruits and vegetables leading to wastage due to inadequate storage facilities.

b. Rates of pay among "cargadores" (product haulers) usually depend on the number of containers that they can carry from one point to another. Hence, they disregard proper handling in their hurried attempts to make more trips.

c. Deterioration during storage occurs because some old stocks are intentionally kept too long in anticipation of eventual price increases;

d. Maintenance of transport, storage and other handling facilities are generally poor in developing countries resulting in a continual source of losses.

e. There is no efficient communication link between producers and wholesalers. Losses will always occur in the absence of a dependable communication system.

Individual fruits and vegetables

The following notes highlight some of the major problems of the more important commodities in the fruit and vegetable group. They are, however, indicative rather than exhaustive.

Bananas and Plantains. Harvesting is generally a one-man operation which frequently results in bruising and abrasions of fruits causing accelerated ripening and consequent decay. Latex staining is prevalent during dehanding. Bunches or hands are often piles one on top of the other without proper protection and fingers are easily detached and oftentimes wasted during transport. This is particularly true of the very open, loose bunches of certain plantain cultivars. In container transport loosely packed hands suffer considerable damage, especially on rough roads. In transit ripening and decay are usually high, notably over long distances.

Mango. Fruits are usually harvested at the time of the day when maximum latex flow is favoured. Latex stain is allowed to dry on the peel, hence immediately reducing consumer acceptability during retail. The collapsibility of the non-rigid crates often used for transport further aggravates quality lose by compression and bruising. The inaccessibility of production area to roads causes serious delays in transport, in addition to a mixed-cargo typo of transport. Stacking in vehicles often does not provide for adequate ventilation. Loading and unloading operations are rather rough. Cold storage is not usually practised. Ripening is mainly aimed at improvement of the appearance for sales purposes and not for maintaining quality.

Papaya. Picking poles injure the fruit and there is a relatively high percentage of fruit dropping on the ground, causing breakage and bruising of ripe fruits. Peduncles are not usually trimmed hence injuring other fruit a within the pack. Rigid containers are not adequately linde, and within a single pack fruits of assorted sizes and maturity stages are often found. In bulk transport fruit e arc piled one on top of the other without any suitable padding materials. High percentage of decay, particularly anthracnose, is the main problem during ripening and in retail.

Citrus. Improper time of harvesting greatly enhances rind injury or oleocellosis. Leaving long stubs on fruit a injures other fruits within the pack. Containers used are large and overpacked, generally without sufficient ventilation. Containers are piled high with the bottom crates bearing the full weight of crates on top. Delays in transport due to poor roads often cause over-ripening or yellowing of the commodity. Poor storage conditions favour decay and physiological disorders such as chilling injury can also occur if cold stores are not wolf managed.

Grapes are attacked by Botrytis, Cladosporium and Alternaria during storage. However, if the storage temperature is strictly maintained between 0° and 2°C, fungal attack can be reduced to a minimum Other loss factors could be berry drop, bruises, injury, water loss, and cracking of berries. Selection of unsuitable container type for packaging of grapes may also lead to heavy transit losses. Transit delays, adverse weather conditions and improper type of carriages, e.g., steel wagons, particularly during hot months may further aggravate transit losses.

Tomato fruits are usually picked when fully ripe, and are therefore very susceptible to cracking, bruising, and consequently decay. Packaging containers often used are deep bamboo crates with insufficient aide reinforcements allowing jarring and compression during transport. Loading and unloading operations are very crude. Handlers tend to throw the pecks rather than lift them gently, on account of their weight. During retail sellers tend to pour the contents of the pack into another container, rather than transferring the fruit gently, thereby increasing bruise damage. Fruit a at the breaker stage are mixed with the fully ripe or three-quarters ripe fruits reducing the market value of the pack. Shrivelling percentage can be high since fruits are often exposed to the sun.

Onions. Insufficient grading is still existent. Spouted, injured and partly decayed bulbs are usually mixed with sound bulbs in a pack. The use of slatted wooden creates is advantageous, especially during transport. Mesh bags of 40-50 kg. capacity are also used. Sacks are thrown rather than lifted, on account of their weight. Packs are piled one on top of the other with no provision for adequate ventilation. Pre-harvest spraying with sprout inhibitors is seldom practised resulting in serious sprouting during storage.

Cabbage/Lettuce. Improper harvesting tools contribute greatly to damage to the produce in the form of out a and abrasions. Trimming of outer leaves is usually not practised. In container transport, large crates are used (50 kg. capacity). Bruising and tearing of the leaves is of common occurrence due to the sharp edges of the containers. Containers are piled one on top of the other with the bottom crates carrying the weight of the heads above. Bulk transport likewise results in higher losses.

Peas and Beans. Factors such as the method of packing, suitability of containers, mode of transport, distance covered, number of transshipments, handling, and storage facilities in the consuming centres, all contribute to the degree of loss reported.

Institutional aspects

The solution to the problems of improper handling of perishable produce in the tropics is rather difficult, owing to the complexity of the problem. It requires the solving of technical problems as well as those of credit, land, transportation and marketing availability. It also requires a change of people's attitudes to proposed solutions and new innovations. Such an approach can only be accomplished over the years, tackling specific problems in a stepwise fashion. There must be concerted efforts by the private and government sectors.

Extension work is needed to show that post-harvest procedures are as important as production techniques. It is not enough to produce good quality commodities through variety improvement and proper regulation of soil and climatic factors. The whole process from planting until the harvested products reach the consuming public must be a mutual undertaking between the growers and those who will handle the product after harvest. Post-harvest handling up to the final marketing stage must be considered as a single system. The success of maintaining the harvest-fresh quality of produce demands control) each atop depending upon the previous one. If the initial quality of the product is poor, no post-harvest treatment can improve it, although careful selection and grading may salvage some good quality produce from a mixed-quality sample. Thus handling procedures from harvesting until the product reaches the consumers are chains of interdependent activities.

The establishment of wholesale markets or cold storage facilities by Governments or other agencies must also meet the general approval of the persons who must use them. Of what use would a modern wholesale market be if the wholesalers would rather use an antiquated one that is more accessible to them? The establishment of a cold storage plant in an area accessible to the users must be accompanied by a sustained information and promotion campaign. Emphasis must be laid on the benefits of the farmer. Realistic rental rates and payments after the sale of product a could attract the farmer to use this facility. Above all' an essential prerequisite to the taking of a decision to erect a new wholesale market is the need to carry out a sound feasibility study to provide suitable information for the decision-making to use.

The establishment of co-operatives has done much in certain countries but only whore the organizations are effectively run and the members are aware of their responsibilities as well as the benefits. The farmers involved in cooperatives have a greater bargaining power and control of the production planning and marketing which may spur then can to improve or maintain the quality of their produce. They may also be encouraged to obtain credit.

Technical knowledge of post-harvest handling has been increasing and there is a need to translate this body of knowledge into systems and techniques that the farmer can understand and use. Agricultural extension information must be reliable. The extension worker must have a, good grasp of his job and must know where to turn when he cannot solve problem himself.

A continuing problem is the low educational lover of farmers in developing countries and their skepticism towards new methods they have never tried or seen before. Farming, especially of fruits and vegetables, is not looked upon as an attractive occupation. Often farming is an occupation of last resort. Vagaries of weather make farming a risky business and knowledgeable and enterprising farmers are few.

In a way, acceptance of improvements and innovation are tied up with the economic progress of a nation. If the buying power of the people is increased, they are more willing to accept the increase in prices associated with the improvement in the post-harvest handling of fruits and vegetables.

References

Pantastico, Er. B. 1980. *FAO/UNEP Expert Consultation on Reduction of Food Losses in Perishable Products of Plant Origin.* (6 May 1980, FAO,, Rome). 54 pp.

Paull, R.E. and J.W. Armstrong (eds). 1994. *Insect pests and fresh horticultural products: treatments and responses.* CAB International, Wallingford, UK, 360 pp.

Sommer, N.F., R.L. Fortlage and D.C. Edwards. 2002. Postharvest Diseases of Selected Commodities. In: Kader, AA (ed). *Postharvest Technology of Horticultural Crops*

Thompson, J.F. 1994. *Ripening facilities. Perishables Handling Newsletter.* University of California, Davis, Special Issue No. 80: 5-8.

13

Importance of Post-harvest Management

Fruits, nuts, and vegetables play a significant role in human nutrition, especially as sources of vitamins, minerals, dietary fibre, and antioxidants. Increased consumption of a variety of fruits and vegetables on a daily basis is highly recommended because of associated health benefits, which include reduced risk of some forms of cancer, heart disease, stroke, and other chronic diseases.

Both quantitative and qualitative losses occur in horticultural commodities between harvest and consumption. Qualitative losses, such as loss in edibility, nutritional quality, caloric value, and consumer acceptability of fresh produce, are much more difficult to assess than are quantitative losses. Quality standards, consumer preferences and purchasing power vary greatly across countries and cultures and these differences influence marketability and the magnitude of post-harvest losses.

Post-harvest losses vary greatly across commodity types, with production areas and the season of production. Losses of fresh fruits and vegetables in developed countries are estimated to range from 2 percent for potatoes to 23 percent for strawberries, with an overall average of 12 percent losses between production and consumption sites. In contrast, the range of produce losses in developing countries varies widely. Losses at the retail, food-service, and consumer levels are estimated at approximately 20 percent in developed countries and about 10 percent in developing countries. Overall, about one third of horticultural crops produced are never consumed by humans.

Reduction of post-harvest losses can increase food availability to the growing world population, decrease the area needed for production, and conserve natural resources. Strategies for loss prevention include:

(1) use of genotypes that have longer post-harvest-life;

(2) use of integrated crop management systems and Good Agricultural Practices that result in good keeping quality of the commodity; and

(3) use of proper post-harvest handling practices in order to maintain the quality and safety of fresh produce.

Although minimizing post-harvest losses of already produced food is more sustainable than increasing production to compensate for these losses, less than 5 percent of the funding of agricultural research and extension programs worldwide is devoted to activities related to maintenance of produce quality and safety during post-harvest handling. This situation must be changed if success is to be achieved in reducing post-harvest losses of horticultural perishables.

Quality Factors

Quality, the degree of excellence or superiority, is a combination of attributes, properties, or characteristics that give each commodity value, in terms of its intended use. The relative importance given to a specific quality attribute varies in accordance with the commodity concerned and with the individual (producer, consumer, and handler) or market concerned with quality assessment.

To producers, high yields, good appearance, ease of harvest, and the ability to withstand long-distance shipping to markets are important quality attributes. Appearance, firmness, and shelf-life are important from the point of view of wholesale and retail marketers. Consumers, on the other hand, judge the quality of fresh fruits, ornamentals, and vegetables on the basis of appearance (including 'freshness') at the time of initial purchase. Subsequent purchases depend upon the consumer's satisfaction in terms of flavor (eating) quality of the edible part of produce. Following is a description of the factors that contribute to the various quality attributes of fresh produce:

Appearance (visual) Quality Factors

These may include size, shape, color, gloss, and freedom from defects and decay. Defects can originate before harvest as a result of damage by insects, diseases, birds, and hail; chemical injuries; and various blemishes (such as scars, scabs, russeting, rind staining). Post-harvest defects may be morphological, physical, physiological, or pathological.

Textural (feel) Quality Factors

These include firmness, crispness, juiciness, mealiness, and toughness, depending on the commodity. Textural quality of horticultural crops is not only important for their eating and cooking quality but also for their shipping ability. Soft fruits cannot be shipped over long distances without substantial losses due to physical injuries. In many cases, the shipment of soft fruits necessitates that they be harvested at less than ideal maturity, from the flavor quality standpoint.

Flavor (eating) Quality Factors

These include sweetness, sourness (acidity), astringency, bitterness, aroma, and off-flavors. Flavor quality involves perception of the tastes and aromas of many compounds. An objective analytical determination of critical components must be coupled with subjective evaluations by a taste panel to yield useful and meaningful information about the flavor quality of fresh fruits and vegetables. This approach can be used to define a minimum level of acceptability. In order to assess consumer preference for the flavor of a given commodity, large-scale testing by a representative sample of consumers is required.

Nutritional Quality Factors

Fresh fruits and vegetables play a significant role in human nutrition, especially as sources of vitamins (Vitamin C, Vitamin A, Vitamin B, thiamine, niacin), minerals, and dietary fibre. Other constituents of fresh fruits and vegetables that may lower the risk of cancer and other diseases include carotenoids, flavonoids, isoflavones, phytosterols, and other phytochemicals (phytonutrients).

Grade standards identify quality attributes in a commodity that are the basis of its use and value. Such standards, if enforced properly, are essential tools of quality assurance during marketing and provide a common language for trade among growers, handlers, processors, and receivers at terminal markets.

Safety Factors

A number of factors threaten the safety of fruits and vegetables. These include naturally-occurring toxicants, such as glycoalkaloids in potatoes; natural contaminants, such as fungal toxins (mycotoxins) and bacterial toxins, and heavy metals (cadmium, lead, mercury); environmental pollutants; pesticide residues; and microbial contamination. While health authorities and scientists regard microbial contamination as the number one safety concern, many consumers rank pesticide residues as their most important safety concern.

Unless fertilized with animal and/or human waste or irrigated with water containing such waste, raw fruits and vegetables should normally be free of most human and animal enteric pathogens. Organic fertilizers, such as chicken manure, should be sterilized prior to their application in fruit and vegetable production, so as to avoid the risk of contaminating fresh produce with Salmonella, Listeria, and other pathogens. Commodities that touch the soil are more likely to be contaminated than those that do not come in contact with the soil. The best approach to achieving and maintaining the safety of fresh fruits and vegetables is to focus on limiting potential contamination during their growth, harvesting, handling, treatment, packaging and storage. Strict adherence to Good Agricultural Practices, i.e. basic food safety principles associated with minimizing biological, chemical and physical hazards from the field throughout the distribution chain of fresh fruits and vegetables; Good Hygienic Practices, i.e. conformance to sanitation and hygienic practices to the extent necessary to protect against contamination of food from direct or indirect sources, is strongly recommended to minimize microbial contamination. Careful handling and washing of all produce to be consumed raw and the strict observance of proper sanitary measures are strongly recommended to reduce microbial contamination at the food-service, retail, and consumer levels.

Factors Influencing Quality and Safety of Horticultural Crops

Genetic Factors

Within each commodity grouping there is a range of genotypic variation in composition, quality, and post-harvest-life potential. Plant breeders have been successful in selecting carrot, sweet potato, and tomato cultivars with comparably high carotenoid levels and vitamin A content; onion and tomato cultivars with longer shelf-lives, sweet corn cultivars that maintain their sweetness longer after harvest; cantaloupe and watermelon cultivars with higher sugar content and firmer flesh, and pineapple cultivars with higher contents of ascorbic acid, carotenoids, and sugars. These are just a few examples of how genetic manipulation has contributed to improving the quality of fruits and vegetables. However, in many cases, commercial cultivars selected for their ability to withstand the rigors of marketing and distribution, tend to lack sufficient sensory quality, in particular flavor.

Horticultural plant breeders have an unprecedented opportunity to address human nutritional needs by developing fruit and vegetable cultivars that are rich in nutrients. In so doing, a multidisciplinary approach should be taken with emphasis the enhancement of nutritional quality for maximum impact on human nutrition and wellness.

Many opportunities exist for applying biotechnology to improving the post-harvest quality and safety of fresh produce. Priority goals in this regard, should, be focused on: (1) attaining and maintaining good flavor and nutritional quality, so as to satisfy consumer demands and (2) introducing resistance to physiological disorders and/or decay-causing pathogens, so as to reduce the use of chemicals on fruits and vegetables.

Climatic Conditions

Climatic factors, in particular temperature and light intensity, greatly impact on the nutritional quality of fruits and vegetables. Consequently, the location of production and the season in which plants are grown can determine their ascorbic acid, carotene, riboflavin, thiamine, and flavonoid contents. In general, the lower the light intensity the lower the ascorbic acid content of plant tissues. Temperature influences the uptake and metabolism of mineral nutrients by plants, since transpiration rates increase with increasing temperature. Rainfall affects water supply to the plant, which may influence the composition of the harvested plant part and its susceptibility to mechanical damage and decay during subsequent harvesting and handling operations.

Cultural Practices

Soil type, the rootstock used for fruit tree cultivation, mulching, irrigation, and fertilization influence the water and nutrient supply to the plant, which can in turn affect the nutritional quality of the harvested plant part. The effect of fertilizers on the vitamin content of plants is less important than are the effects of genotype and climatic conditions. The effects of mineral and elemental uptake from fertilizers by plants are, however, significant and variable. Selenium and sulfur uptake for example influence the concentrations of organosulfur compounds in Allium and Brassica species. High calcium uptake in fruits has been shown to reduce respiration rates, and ethylene production, to delay ripening, increase firmness, and reduce the incidence of physiological disorders and decay, all of which result in increased post-harvest shelf-life. High nitrogen content on the other hand, is often associated with reduced post-harvest-life due to increased susceptibility to mechanical damage, physiological disorders, and decay. Increasing the nitrogen and/or phosphorus supply to citrus trees results in somewhat reduced acidity and ascorbic acid content in citrus fruits, while increased potassium fertilization results in increased acidity and ascorbic acid content.

Numerous physiological disorders are associated with mineral deficiencies. Bitter pit of apples; blossom-end rot of tomatoes, peppers, and watermelons; cork spot in apples and pears; and red blotch of lemons are all associated with calcium deficiency in these fruits. Boron deficiency results in corking of apples, apricots, and pears; lumpy rind of citrus fruits, and cracking of apricots. Poor color of stone fruits may be related to iron and/or zinc deficiencies. Excess sodium and/or chloride (due to salinity) results in reduced fruit size and higher soluble solids content.

Severe water stress results in increased sunburn of fruits, irregular ripening of pears, tough and leathery texture in peaches, and incomplete kernel development in nuts. Moderate water stress reduces fruit size and increases soluble solids content, acidity, and ascorbic acid content. On the other hand, excess water supply to plants results in cracking of fruits (such as cherries, plums, and tomatoes), excessive turgidity leading to increased susceptibility to physical damage (such as oil spotting on citrus fruits), reduced firmness, delayed maturity, and reduced soluble solids content.

Cultural practices such as pruning and thinning determine the crop load and fruit size, which can in turn influence the nutritional composition of fruit. The use of pesticides and growth regulators does not directly influence fruit composition but may indirectly affect it due to delayed or accelerated fruit maturity. Effective pre-harvest disease control greatly influences disease incidence and severity during post-harvest handling of fruits and vegetables

Maturity at Harvest in Relation to Quality

Maturity at harvest is the most important determinant of storage-life and final fruit quality. Immature fruit are highly susceptible to shriveling and mechanical damage, and are of inferior flavor quality when ripe. Overripe fruit are likely to become soft and mealy with insipid flavor soon after harvest. Fruit picked either prematurely or too late, are more susceptible to post-harvest physiological disorders than are fruit picked at the proper stage of maturity.

With a few exceptions all fruits attain optimal eating quality when allowed to ripen on the plant. Some fruits are, however, picked at a mature but unripe stage of development so as to allow them to withstand post-harvest handling conditions when shipped over long-distances. Maturity indices for such fruit are based on a compromise between those indices that would ensure the best eating quality to the consumer and those that provide flexibility in marketing.

Fruit can be divided into two groups: (1) those that are incapable of continuing their ripening process once removed from the plant, and (2) those that can be harvested at the mature stage and allowed to ripen off the plant. Group 1 includes cane berries, cherry, citrus fruits, grape, lychee, pineapple, pomegranate, strawberry, and tamarillo. Group 2 on the other hand, includes apple, apricot, avocado, banana, cherimoya, guava, kiwifruit, mango, nectarine, papaya, passion fruit, pear, peach, persimmon, plum, quince, sapodilla, and sapote.

Fruit of the Group 1 category, produce very small quantities of ethylene and do not respond to ethylene treatment except in terms of degreening (removal of chlorophyll); these should be picked when fully-ripe, if good flavor quality is to be ensured. Fruit of the Group 2 category on the other hand, produce comparably larger quantities of ethylene which is associated with their ripening, and undergo more rapid and uniform ripening upon exposure to ethylene.Many vegetables, in particular leafy vegetables, and immature fruit-vegetables, attain optimum eating-quality prior to reaching full maturity. This often results in delayed harvest, and consequently in produce of low quality.

Method of harvesting in relation to physical damage and uniformity of maturity

The method of harvesting (hand vs mechanical) can significantly impact upon the composition and post-harvest quality of fruits and vegetables. Mechanical injuries (such as bruising, surface abrasions and cuts) can accelerate loss of water and vitamin C resulting in increased susceptibility to decay-causing pathogens. Most fresh fruits and vegetables and all flowers are harvested by hand. Root crops (such as carrot, onion, potato, and sweet potato) and some commodities destined for processing (such as processing tomatoes, European plums, and tree nut crops) are mechanically harvested.

Management of harvesting operations, whether manual or mechanical, can have a major impact on the quality of harvested fruits and vegetables. Proper management procedures include selection of optimum time to harvest in relation to product maturity and climatic conditions, training and supervision of workers, and proper implementation of effective quality control. Expedited and careful handling, immediate cooling after harvest, maintenance of optimum temperatures during transit and storage, and effective decay-control procedures are important factors in the successful post-harvest handling of fruits and vegetables.

Attention must be paid to all of these factors, regardless of the method of harvesting used. These factors are nevertheless more critical in the case of mechanically harvested commodities.

It should be noted that any practice that reduces the number of produce handling steps will help minimize losses. Field packing (selection, sorting, trimming, and packaging) of produce at the time of harvest can greatly reduce the number of handling steps in preparation for marketing. Mobile field packing stations with adequate shading are used for those fruits (such as grapes and strawberries) and vegetables (such as broccoli, cauliflower, and green beans) that do not require washing as part of their preparation for marketing.

Procedures Critical to Maintaining the Quality and Safety

Packing and Packaging

Preparation of produce for market may be done either in the field or at the packing house. This involves cleaning, sanitizing, and sorting according to quality and size, waxing and, where appropriate, treatment with an approved fungicide prior to packing into shipping containers. Packaging protects the produce from mechanical injury, and contamination during marketing. Corrugated fiberboard containers are commonly used for the packaging of produce, although reusable plastic containers can be used for that purpose. Packaging accessories such as trays, cups, wraps, liners, and pads may be used to help immobilize the produce within the packaging container while serving the purpose of facilitating moisture retention, chemical treatment and ethylene absorption. Either hand-packing or mechanical packing systems may be used. Packing and packaging methods can greatly influence air flow rates around the commodity, thereby affecting temperature and relative humidity management of produce while in storage or in transit.

Temperature and Relative Humidity Management

Temperature is the most important environmental factor that influences the deterioration of harvested commodities. Most perishable horticultural commodities have an optimal shelf-life at temperatures of approximately 0 °C. The rate of deterioration of perishables however increases two to three-fold with every 10 °C increase in temperature Temperature has a significant effect on how other internal and external factors influence the commodity, and dramatically affects spore germination and the growth of pathogens. Temperatures either above or below the optimal range for fresh produce can cause rapid deterioration due to the following disorders:

Freezing Perishable commodities are generally high in water content, and possess large, highly vacuolate cells. The freezing point of their tissues is relatively high (ranging from -3 °C to -0.5 °C), and disruption caused by freezing generally results in immediate collapse of their tissues and a total loss of cellular integrity. Freezing occurs in cold storage systems either due to inadequate refrigerator design, or to thermostat failure. Freezing can also occur upon exposure to inclement weather conditions as occurs when produce is allowed to remain for even short periods of time on unprotected transportation docks during winter.

Chilling injury Some commodities (chiefly those native to the tropics and subtropics) respond unfavorably to storage at low temperatures which are well above their freezing points, but below a critical temperature termed their chilling threshold temperature or lowest safe temperature (Table 1). Chilling injury is manifested in a variety of symptoms including surface and internal discoloration, pitting, water soaking, failure to ripen, uneven ripening, development of off flavors and heightened susceptibility to pathogen attack.

Table 1: Classification of chilling-sensitive fruits and vegetables according to their lowest safe temperature for transport and storage

Lowest safe temperature (°C)	*Commodity*
3	Asparagus, cranberry, jujube
4	Cantaloupe, certain apple cultivars (such as McIntosh and Yellow Newton), certain avocado cultivars (such as Booth and Lula), lychee, potato, tamarillo
5	Cactus pear, cowpeas, durian, feijoa, guava, kumquat, lima bean, longan, mandarin, orange, pepino
7	Certain avocado cultivars (such as Fuerte and Hass), chayote, okra, olive, pepper, pineapple, pomegranate, snap bean
10	Carambola, cucumber, eggplant, grapefruit, lime, mango (ripe), melons (casaba, crenshaw, honeydew, persian), papaya, passion fruit, plantain, rambutan, squash (soft rind), taro, tomato (ripe), watermelon
13	Banana, breadfruit, cherimoya, ginger, jackfruits, jicama, lemon, mango (mature-green), mangosteen, pumpkin and hard-rind squash, sapotes, sweet potato, tomato (mature-green), yam

Heat injury High temperature conditions are also injurious to perishable crops. Transpiration is vital to maintaining optimal growth temperatures in growing plants. Organs removed from the plant, however, lack the protective effects of transpiration, and direct sources of heat, such as sunlight, can rapidly elevate the temperature of tissues to above the thermal death point of their cells, leading to localized bleaching, necrosis or general collapse.

Chilling Injury on Mangoes

Relative humidity (RH) is defined as the moisture content (as water vapor) of the atmosphere, expressed as a percentage of the amount of moisture that can be retained by the atmosphere (moisture holding capacity) at a given temperature and pressure without condensation. The moisture holding capacity of air increases with temperature. Water loss is directly proportional to the vapor pressure difference (VPD) between the commodity and its environment. VPD is inversely related to the RH of the air surrounding the commodity.

RH can influence water loss, decay development, the incidence of some physiological disorders, and uniformity of fruit ripening. Condensation of moisture on the commodity (sweating) over long periods of time is probably more important in enhancing decay than is the RH of ambient air. An appropriate RH range for storage of fruits is 85 to 95 percent while that for most vegetables varies between 90 and 98 percent. The optimal RH range for dry onions and pumpkins is 70 to 75 percent. Some root vegetables, such as carrot, parsnip, and radish, can best be held at 95 to 100 percent RH.

RH can be controlled by one or more of the following procedures:

(1) adding moisture (water mist or spray, steam) to air with the use of humidifiers;

(2) regulating air movement and ventilation in relation to the produce load in the cold storage room;

(3) maintaining the temperature of the refrigeration coils in the storage room or transit vehicle to within about 1 °C of the air temperature;

(4) providing moisture barriers that insulate walls of storage rooms and transit vehicles;

(5) adding polyethylene liners in packing containers and using perforated polymeric films for packaging;

(6) wetting floors in storage rooms;

(7) adding crushed ice in shipping containers or in retail displays for commodities that are not injured by the practice;

(8) sprinkling produce with sanitized, clean water during retail marketing of commodities that benefit from misting, such as leafy vegetables, cool-season root vegetables, and immature fruit vegetables (such as snap beans, peas, sweet corn, and summer squash).

Cooling Methods

Temperature management is the most effective tool for extending the shelf life of fresh horticultural commodities. It begins with the rapid removal of field heat by using one of the cooling methods listed in Table 2.

Table 2: Comparison of methods used for cooling

Variable	*Cooling method*				
	Ice	*Hydro*	*Vacuum*	*Forced-air*	*Room*
Cooling times (h)	0.1-0.3	0.1-1.0	0.3-2.0	1.0-10.0	20-100
Water contact with product	yes	yes	no	no	no
Product moisture loss (%)	0-0.5	0-0.5	2.0-4.0	0.1-2.0	0.1-2.0
Capital cost	high	low	medium	low	low
Energy efficiency	low	high	high	low	low

Packing fresh produce with crushed or flaked ice provides rapid cooling, and can provide a source of cooling and high RH during subsequent handling. The use of crushed ice is, however, limited to produce that is tolerant to direct contact with ice and packaged in moisture-resistant containers.

Clean, sanitized water is used as the cooling medium for the hydrocooling (shower or immersion systems) of commodities that tolerate water contact and are packaged in moisture-resistant containers. Vacuum cooling is generally applied to leafy vegetables that release water vapor quickly, thereby allowing them to be rapidly cooled. During forced-air cooling on the other hand, refrigerated air is forced through produce packed in boxes or pallet bins. Forced-air cooling is applicable to most horticultural perishables.

Precise temperature and RH management are required to provide the optimum environment for fresh fruits and vegetables during cooling and storage. Precision temperature management (PTM) tools, including time-temperature monitors, are increasingly being employed in cooling and storage facilities.

Refrigerated Transport and Storage

Cold storage facilities should be appropriately designed, of good construction, and be adequately equipped. Their insulation should include a complete vapor barrier on the warm side of the insulation; sturdy floors; adequate and well-positioned doors for loading and unloading; effective distribution of refrigerated air; sensitive and properly located controls; refrigerated coil surfaces designed to adequately minimize differences between the coil and air temperatures; and adequate capacity for expected needs. Commodities should be stacked in the cold room or the refrigerated vehicle with air spaces between pallets and room walls so as to ensure proper air circulation. Storage rooms should not be loaded beyond their capacity limit if proper cooling is to be achieved. Commodity temperature rather than air temperature should be measured in these facilities.

Temperature management during transportation of fresh fruits and vegetables over long distances is critical. Loads must be stacked so as to enable proper air circulation, in order to facilitate removal of heat from the produce as well as to dissipate incoming heat from the atmosphere and off the road. Stacking of loads must also incorporate consideration for minimizing mechanical damage. Transit vehicles must be cooled prior to loading the fresh produce. Delays between cooling after harvest and loading into transit vehicles should also be avoided. Proper temperature maintenance should be ensured throughout the handling system.

As far as possible, environmental conditions (temperature; relative humidity; concentrations of oxygen, carbon dioxide, and ethylene) should be optimized in transport vehicles. Treatment with ethylene to initiate ripening during transportation is feasible, and is commercially used to a limited extent on mature green bananas and tomatoes. Produce should be cooled prior to loading and should be loaded with an air space between the palletized product and the walls of the transport vehicles in order to facilitate temperature control. Vibration during transportation should be minimized, so as to avoid damage due to bruising. Controlled-atmosphere and precision temperature management should, where possible, be observed so as to allow non-chemical insect control for markets which possess quarantine restrictions against pests endemic to exporting countries and for markets that do not want their produce exposed to chemical fumigants.

Mixing several produce items in one load is common and often compromises have to be made in selecting an optimal temperature and atmospheric composition when transporting chilling-sensitive with non-chilling sensitive commodities or ethylene-producing with ethylene-sensitive commodities. In the latter case, ethylene scrubbers can be used to remove ethylene from the circulating air within the vehicle. Several types of insulating pallet covers are available for protecting chilling-sensitive commodities when transported with non-chilling-sensitive commodities at temperatures below their threshold chilling temperatures.

Cold Chain and Its Importance

The cold chain encompasses all the critical steps and processes that foods and other perishable products must undergo in order to maintain their quality. Like any chain, the cold chain is only as strong as its weakest link. Major limitations experienced by the cold-chain include poor temperature management due to either the lack of, or limitations in, refrigeration, handling, storage, and humidity control. Investment in cold chain infrastructure ultimately leads to a reduction in the level of losses and quality degradation in fresh produce, with overall net positive economic returns.

Return on Investment in Temperature

Deficiencies in cold chain management whether due to limitations in refrigeration, improper handling and storage, or inadequate humidity control, can lead to losses in profits as well as in horticultural crops. Overcoming such deficiencies necessitates improvements in methodologies, operations and handling along the chain. Often the level of investment required in overcoming such deficiencies is minimal in comparison to the level of losses sustained over time.

A University of California study determined that a one-hour delay in cooling strawberries after harvest resulted in a 10 percent loss due to decay during marketing. The resulting economic loss exceeded the increased cost of expedited handling of the strawberries by more frequent deliveries of harvested fruit to the cooling facility and initiation of forced-air cooling. Similarly, a University of Georgia study showed that poor temperature management of lettuce resulted in a net income loss of US$172.50 per truck-load of 900 cartons.

A University of California study determined that excess weight loss coupled with color deterioration resulting from delays between harvest and cooling, improper refrigeration temperature and relative humidity control during the shipping of table grapes resulted in a 15 percent loss in the value of that commodity. Resultant monetary losses were greater than the cost of improved management of temperature and RH of the grapes with perforated plastic liners in the boxes and by minimizing delays prior to cooling with humidified, forced air.

Treatments Designed to Reduce Produce Contamination

Treatments to Reduce microbial contamination

Over the past few years, food safety has become and continues to be the number one concern of the fresh produce industry. A "Guide to Minimize Microbial Food Safety Hazards for Fresh Fruits and Vegetables," was published by the U.S. Food and Drug Administration in October 1998. This guide is based on the following principles:

(1) Prevention of microbial contamination of fresh produce is favored over reliance on corrective actions once contamination has occurred;

(2) In order to minimize microbial food safety hazards in fresh produce, growers, packers, or shippers should use good agricultural and management practices in those areas over which they have control;

(3) Fresh produce can become microbiologically contaminated at any point along the farm-to-table food chain. Human and/or animal faeces are the source of microbial contamination of fresh produce;

(4) Whenever water comes in contact with produce, its quality dictates the potential for contamination. The potential of microbial contamination from water used with fresh fruits and vegetables must be minimized;

(5) The use of animal manure or municipal biosolid wastes as fertilizers should be closely managed in order to minimize the potential for microbial contamination of fresh produce; and

(6) Worker hygiene and sanitation practices during production, harvest, sorting, packing, and transport play a critical role in minimizing the potential for microbial contamination of fresh produce."

A training manual for trainers, entitled "Improving the Safety and Quality of Fresh Fruits and Vegetables," was published by the United States Food and Drug Administration (USFDA) in November 2002, with the objective of providing uniform, broad-based scientific and practical information on the safe production, handling, storage, and transport of fresh produce.

Clean water containing an appropriate concentration of sanitizers is required in order to minimize the potential transmission of pathogens from water to produce, from healthy to infected produce within a single lot, and from one lot of produce to another, over time. Waterborne microorganisms, including post-harvest plant pathogens and agents of human illness, can be rapidly acquired and taken up on plant surfaces. Natural plant surface contours, natural openings, harvest and trimming wounds and scuffing can provide points of entry as well as safe harbor for microbes. When located in these protected sites, microbes are largely unaffected by common or permitted doses of post-harvest water sanitizing treatments. It is therefore essential that the sanitizer concentration is sufficient to kill microbes before they attach or become internalized in produce. The concentration of sanitizer is important in some pre-harvest water uses (such as spraying pesticides or growth regulators) and in all post-harvest procedures involving water, including washing, cooling, water-mediated transport (flumes), and post-harvest drenching with calcium chloride or other chemicals.

Treatments to Minimize Water Loss

Transpiration, or evaporation of water from the plant tissues, is one of the major causes of deterioration in fresh horticultural crops after harvest. Water loss through transpiration not only results in direct quantitative losses (loss of saleable weight), but also causes losses in appearance (wilting, shriveling), textural quality (softening, flaccidity, limpness, loss of crispness and juiciness), and nutritional quality. Transpiration can be controlled either through the direct application of post-harvest treatments to the produce (surface coatings and other moisture barriers) or through manipulation of the environment (maintenance of high relative humidity).

Treatments that can be applied to minimize water loss in fruits and vegetables include:

a. Curing of certain root vegetables, such as garlic, onion, potato, and sweet potato.

b. Waxing and the use of other surface coatings on commodities, such as apple, citrus fruits, nectarine, peach, plum, pomegranate, and tomato.

c. Packaging in polymeric films that act as moisture barriers.

d. Careful handling to avoid physical injuries, which increase water loss from produce.

e. Addition of water to those commodities that tolerate misting with water, such as leafy vegetables.

Treatments to Reduce Ethylene Damage

The promotion of senescence in harvested horticultural crops by ethylene (1 ppm or higher) results in acceleration of deterioration and reduced post-harvest life. Ethylene accelerates chlorophyll degradation and induces yellowing of green tissues, thus reducing the quality of leafy, floral, and immature fruit-vegetables and foliage ornamentals. Ethylene induces abscission of leaves and flowers, softening of fruits, and several physiological disorders. Ethylene may increase decay development of some fruits by accelerating their senescence and softening and by inhibiting the formation of antifungal compounds in the host tissue. In some cases, ethylene may stimulate the growth of fungi, such as Botrytis cineria on strawberries and Penicillium italicum on oranges.

The incidence and severity of ethylene-induced deterioration symptoms is dependent upon temperature, time of exposure, and ethylene concentration. The yellowing of cucumbers can, for example, result from exposure to either 1 ppm ethylene over 2 days or to 5 ppm ethylene over 1/2 day at 10 °C. Ethylene effects are cumulative throughout the post-harvest life of the commodity.

Treatment of ornamental crops with 1-methylcyclopropene (1-MCP), which is an ethylene action inhibitor, provides protection against ethylene damage. The commercial use of this product at concentrations of up to 1 ppm on apples, apricots, avocados, kiwifruit, mangoes, nectarines, papayas, peaches, pears, persimmons, plums, and tomatoes was approved by the United States Environmental Protection Agency. The use of 1-MCP will no doubt be extended to several other fruits and vegetables, and to use in other regions.

Treatments for Decay Control

A major cause of losses in perishable crops is the action of a number of microorganisms on the commodity. Fungi and bacteria may infect the plant organ at any time. "Latent" infections, in which fungi invade fruit tissues shortly after flowering, become apparent only at the onset of ripening. Post-harvest rots frequently occur as a result of rough handling during the marketing process and are caused by a wide array of microorganisms. The grey mold Botrytis cineria is a very important cause of loss in many commodities (such as grapes, kiwifruit, pomegranates, raspberries, and strawberries), and is an aggressive pathogen, even at low temperatures. Virus infection frequently lowers the quality of perishable commodities, usually as a result of visual deterioration, although viruses may also affect flavor and composition.

Sanitation practices include treatment to reduce populations of microorganisms on equipment, on the commodity, and in the wash water used to clean it. Water washes alone are effective in removing nutrients that allow microorganisms to grow on the surfaces of produce as well as in removing inoculum of post-harvest pathogens. The addition of sanitizers to water dumps and spray or dip washes, reduces inoculum levels of decay-causing organisms from fruit surfaces, inactivates spores brought into solution from fruit or soil and prevents the secondary spread of inoculum in water. Treatments for decay control include: (1) heat treatments, such as dipping mangoes in water at a temperature of 50 °C, for 5 minutes in order to reduce subsequent development of anthracnose; (2) use of post-harvest fungicides, such as imazalil and/or thiabendazole on citrus fruits; (3) use of biological control agents, such as "Bio-Save" (Pseudomonas syringae) and "Aspire" (Candida oleophila) alone or in combination with fungicides at lower concentrations on citrus fruits; (4) use of growth regulators such as gibberellic acid or 2, 4-D to delay senescence of citrus fruits; (5) use of 15-20 percent CO_2 in air or 5 percent O_2 on strawberries, cane berries, figs, and pomegranates; and (6) use of SO_2 fumigation (100 ppm for one hour) on grapes.

Table 3: Conditions for curing root, tuber, and bulb crops

Crops	*Temperature(°C)*	*Relative humidity%*	*Duration(days)*
Cassava root	30-40	90-95	2-5
Bulbs	30-45	60-75	1-4
Potato tubers	15-20	85-90	5-10
Sweet Potato roots	30-32	85-90	4-7
Yam tubers	32-40	90-100	1-4

Treatments for Insect Control

Fresh fruits, vegetables and flowers may harbor a large number of insects during post-harvest handling. Many of these insect species, in particular fruit flies of the family Tephritidae (e.g. Mediterranean fruit fly, Oriental fruit fly, Mexican fruit fly, Caribbean fruit fly), can seriously disrupt trade among countries. The identification and application of acceptable disinfestation treatments including irradiation will greatly facilitate globalization of trade in fresh produce. Criteria for the selection of the most appropriate disinfestation treatment for a specific commodity include cost, the efficacy of that treatment against insects of concern, safety of the treatment as well as the ability of that treatment to preserve and maintain produce quality. Currently approved quarantine treatments, other than irradiation, include certification of insect-free areas, use of chemicals (e.g. methyl bromide, phosphine, hydrogen cyanide), cold treatments, heat treatments, and combinations of these treatments, such as methyl bromide fumigation in conjunction with cold treatment. The use of alternative treatments, such as fumigants (carbonyl sulphide, methyl iodide, sulphuryl fluoride) and insecticidal atmospheres (oxygen concentrations of less than 0.5 percent and/or carbon dioxide concentrations ranging between 40 and 60 percent) alone or in combination with heat treatments, and ultraviolet radiation, are currently under investigation. These treatments are not, however, broad-spectrum treatments and are potentially phytotoxic when applied to some commodities.

Most insects become sterile when subjected to irradiation doses ranging between 50 and 750 Gy. The actual dosage required to produce sterility in insects varies in accordance with the species concerned and its stage of development. An irradiation dose of 250 Gy was approved by the United States quarantine authorities for application to fresh commodities, such as lychees, mangoes, and papayas in light of the efficacy of that dose in preventing the reproduction of tropical fruit flies. Irradiation doses of 250 Gy can be tolerated by most fresh fruits and vegetables with minimal detrimental effects on quality. Doses ranging between 250 and 1000 G2, can, however, be damaging to some commodities. Fruits, in general, are more tolerant to the expected dose range (250 to 500 Gy absorbed by fruits on the inside vs. those on the outside of the pallet) than non-fruit vegetables and cut flowers. Detrimental effects of irradiation on fresh produce may include loss of green color (yellowing), abscission of leaves and petals, tissue discoloration, and uneven ripening. These detrimental effects may not become visible until after the commodity reaches the market.

Treatments Designed to Manipulate the Environment

Modified atmosphere storage

When used as supplements to keeping fresh horticultural perishables within their optimum ranges of temperature and relative humidity, controlled atmospheres (CA) or modified atmospheres (MA) can serve to extend their post-harvest-life (Table 4). Optimum oxygen and carbon dioxide concentrations lower respiration and ethylene production rates, reduce ethylene action, delay ripening and senescence, retard the growth of decay-causing pathogens, and control insects. CA conditions which are not suited to a given commodity can, however, induce physiological disorders and enhance susceptibility to decay.

Table 4: Classification of horticultural crops according to their controlled atmosphere storage potential at optimum temperatures and relative humidities

Range of storageduration (months)	*Commodity*
More than 12	Almond, Brazil nut, cashew, filbert, macadamia, pecan, pistachio, walnut, dried fruits and vegetables
6-12	Some cultivars of apples and European pears
3-6	Cabbage, Chinese cabbage, kiwifruit, persimmon, pomegranate, some cultivars of Asian pears
1-3	Avocado, banana, cherry, grape (no SO_2), mango, olive, onion (sweet cultivars), some cultivars of nectarine, peach and plum, tomato (mature-green)
<1	Asparagus, broccoli, cane berries, fig, lettuce, muskmelons, papaya, pineapple, strawberry, sweet corn; fresh-cut fruits and vegetables; some cut flowers.

Several refinements in CA storage technology have been made in recent years. These include: the creation of nitrogen-on-demand by separation of nitrogen from compressed air through the use of either molecular sieve beds or membrane systems; use of low (0.7 to 1.5 percent) oxygen concentrations that can be accurately monitored and controlled; rapid establishment of CA, ethylene-free CA, programmed (or sequential) CA (such as storage in 1 percent O_2 for 2 to 6 weeks followed by storage in 2-3 percent O_2 for remainder of the storage period), and dynamic CA where levels of O_2 and CO_2 are modified as needed based on monitoring specific attributes of produce quality, such as ethanol concentration and chlorophyll fluorescence.

The use of CA in refrigerated marine containers continues to benefit from technological and scientific developments. CA transport is used to continue the CA chain for commodities (such as apples, pears, and kiwifruits) that had been stored in CA immediately after harvest. CA transport of bananas permits their harvest at a more advanced stage of maturity, resulting in the attainment of higher yields at the field level. In the case of avocados, CA transport facilitates use of a lower shipping temperature (5 °C) than if shipped in air, since CA ameliorates chilling injury symptoms. CA in combination with precision temperature management allows insect control without the use of chemicals, in commodities destined for markets that have restrictions against pests endemic to exporting countries and for markets with a preference for organic produce.

The use of polymeric films for packaging produce and their application in modified atmosphere packaging (MAP) systems at the pallet, shipping container (plastic liner), and consumer package levels continues to increase. MAP(usually designed to maintain 2 to 5 percent O_2levels and 8 to 12 percent CO_2 levels) is widely applied in extending the shelf-life of fresh-cut fruits and vegetables. Use of absorbers of ethylene, carbon dioxide, oxygen, and/or water vapor as part of MAP is increasing. Although much research has been done on the use of surface coatings to modify the internal atmosphere within the commodity, commercial applications are still very limited due to inherent biological variability of commodities.

At the commercial level, CA is most widely applied during the storage and transport of apples and pears. It is also applied to a lesser extent on kiwifruits, avocados, persimmons, pomegranates, nuts and dried fruits. Atmospheric modification during long-distance transport is used for apples, avocados, bananas, blueberries, cherries, figs, kiwi-fruits, mangoes, nectarines, peaches, pears, plums, raspberries and strawberries. Technological developments geared toward providing CA during transport and storage at reasonable cost (positive benefit/cost ratio) are essential if the application of this technology to fresh fruits and vegetables is to be expanded.

Although MA and CA have both been shown to be effective in extending the post-harvest life of many commodities, their commercial application has been limited by the relatively high cost of these technologies. There are however a few cases in which a positive return on investment (cost/benefit ratio) can be demonstrated. In a comparison of losses due to decay during retail marketing of strawberries shipped in air and those shipped in an environment consisting of 15 percent CO_2-enriched air (modified atmosphere within pallet cover), the use of modified atmosphere was observed to reduce losses by 50 percent (an average of 20 percent losses was sustained in strawberries shipped in air vs 10 percent losses in those shipped by MA). The economic loss of 10 percent value (US$50-75 per pallet) was much greater than the cost of using MA (US$15-25 per pallet).

Use of controlled atmosphere (CA) during marine transportation can extend the post-harvest-life of those fruits and vegetables that would normally have a short post-harvest-life potential, thereby allowing the use of marine transportation instead of air transport for the shipment of such produce. In terms of cost and benefit, savings realized with the use of marine transportation are much greater than is the added cost of CA service

Ethylene Exclusion and Removal

Many green vegetables and most horticultural produce are quite sensitive to ethylene damage. Their exposure to ethylene must therefore be minimised. Ethylene contamination in ripening rooms can be minimized by 1) using ethylene levels of 100 ppm instead of the higher levels often used in commercial ripening operations, 2) venting ripening rooms to the outside on completion of exposure to ethylene, 3) at least once per day ventilating the area around the ripening rooms or installing an ethylene scrubber, 4) use of battery-powered forklifts instead of engine-driven units in ripening areas.

Ethylene-producing commodities should not be mixed with ethylene-sensitive commodities during storage and transport. Potassium permanganate, an effective oxidizer of ethylene, is commercially used as an ethylene scrubber. Scrubbing units based on the catalytic oxidation of ethylene are used to a limited extent in some commercial storage facilities.

Return on Investment in Reducing Ethylene Damage

A University of California study showed that the use of an ethylene scrubber in storage facilities used for lettuce significantly reduced russet spotting. The difference in value of lettuce that was protected from ethylene vs that which was exposed to ethylene was estimated to be 20 to 25 percent, which was greater than the cost of the ethylene scrubber. Similar results were found with kiwi fruits, which soften very rapidly when exposed to ethylene levels as low as 50 ppb.

Treatments to Enhance Uniformity in Fruit Ripening

Ethylene treatment is commercially used to enhance the rate and uniformity of ripening of fruits such as bananas, avocados, mangoes, tomatoes, and kiwifruits. Optimal ripening conditions are as follows:

Temperature:	18 °C to 25 °C (65 °C to 77 °F)
Relative humidity:	90 to 95 percent
Ethylene concentration:	10 to 100 ppm
Duration of treatment:	24 to 74 hours depending on fruit type and stage of maturity
Air circulation:	Sufficient to ensure distribution of ethylene within the ripening room
Ventilation:	Require adequate air exchange in order to prevent accumulation of O_2 which reduces the effectiveness of C_2H_4.

Selection of Appropriate Post-harvest Technologies

The basic recommendations for maintaining post-harvest quality and safety of produce are the same regardless of the distribution system (direct marketing, local marketing, export marketing). However, the level of technology needed to provide the recommended conditions varies in accordance with the distance and time between production and consumption sites, the intended use of the produce (fresh vs. processing) and the target market. In situations where the point of sale is only a matter of hours away from the site of harvest, careful harvesting and handling and the observance of proper sanitation practices are adequate measures for assuring the quality and safety of fruits and vegetables targeted for the fresh market. Pre-cooling, refrigeration and packaging however become essential when fresh produce must be moved over long distances. The following should be considered when selecting appropriate post-harvest technologies:

a. The technology used elsewhere is not necessarily the best for use under conditions of a given developing country. Many of the recent developments in post-harvest technology in developed countries have come about in response to the need to economize on labor, materials, and energy use, and to protect the environment. Currently used practices in other countries should be studied, but only those which are appropriate for local conditions should be adopted and used.

b. Expensive equipment and facilities are useless without proper management. Furthermore, over-investment in handling facilities can result in economic losses, if consumers in the target market are unable to absorb these added costs. Proper education of all stakeholders along the post-harvest chain (growers, harvesters, handlers and those involved in marketing) is more critical than the level of sophistication of the equipment used in post-harvest handling. Effective training and supervision of personnel must, therefore, be an integral part of quality and safety-assurance programs.

c. Commodity requirements can be met through the use of simple and inexpensive methods in many cases. Proper temperature management procedures, for example, include: (1) Protection from exposure to the sun; (2) Harvesting during cooler periods of the day or even at night; (3) Adequate ventilation in containers and non-refrigerated transport vehicles; (4) Use of simple and inexpensive cooling procedures, such as evaporative cooling or night ambient air; and (5) Expedited handling of fresh produce.

d. Mechanical injuries are major causes of losses in the quality and quantity of fresh horticultural commodities in all handling systems. The incidence and severity of mechanical injury can be greatly minimized by reducing the number of steps involved in harvesting and handling and by educating all personnel involved, about the need for careful handling.

e. Assuring food safety throughout the post-harvest handling system is very critical to successful marketing of produce and should be given the highest priority.

f. Solving the post-harvest technology problems in a given country necessitates cooperation and effective communication among research and extension personnel. Post-harvest horticulturists therefore need to coordinate their efforts and to cooperate with production horticulturists, agricultural marketing economists, engineers, food technologists, microbiologists, and others who may be involved in various aspects of the marketing systems. In most cases, solutions to existing problems in the post-harvest handling system require the use of existing information rather than new research. The following is a proposed program for improving the post-harvest handling system in a developing country:

    i) Survey the magnitude and causes of losses in quality and quantity during harvesting and post-harvest handling of commodities.

    ii) Survey available tools and facilities for harvesting, packing and packaging, transport, storage, and marketing of each commodity in the region of production and during the season of production.

iii) Evaluate the impact of simple modifications in the handling system (such as stage of harvesting, method of harvest, type of container, and quality sorting) on quality and safety maintenance.

iv) Extend information on recommended harvesting and handling procedures to all those who can use it. All appropriate extension methods for the intended audiences should be used.

v) Identify problems which require further research, conduct the needed research and extend any new information when completed to those who can use it.

References

Calderon, M. and R. Barkai-Golan, eds. 1990. *Food preservation by modified atmospheres*. Boca Raton, FL: CRC Press, 402 pp.

Food and Agriculture Organization of the United Nations. 1981. *Food loss prevention in perishable crops*. Rome: FAO Agricultural Series Bulletin 43, 72 pp.

Shewfelt, R.L. and S.E. Prussia, eds. 1993. *Post-harvest handling: A systems approach.* San Diego, CA: Academic Press, 358 pp.

Thompson, A.K. 1996. *Post-harvest technology of fruits and vegetables*. Oxford: Blackwell Science, 410 pp.

Wills, R., B. McGlasson, D. Graham, and D. Joyce. 1998. *Postharvest: An introduction to the physiology and handling of fruit, vegetables and ornamentals.* Wallingford: CAB International, 262 pp.

Bibliography

ASEAN-PHTRC. 1984. *Village Level Handling of Fruits and Vegetables: Traditional Practices and Technological Innovations.* Postharvest Horticulture and Training Center, College of Agriculture, University of the Philippines at Los Baños. Extension Bulletin No. 1

Bachmann, J. and R. Earles. 2000. Postharvest Handling of Fruits and Vegetables. *ATTRA Horticulture Technical* Note. 19 pp. (http://www.attra.ncat.org)

Bourne, M.C. 1977. *Post-harvest Food Losses - The Neglected Dimension in Increasing the World Food Supply*. New York, Cornell University International Agriculture.

Boxall, R.A. 1986. *A Critical Review of the Methodology for Assessing Farm-level Grain Losses after Harvest.* Londres, TDRI, G191, 139 p.

Brackett, R.E. 1999. Incidence, contributing factors, and control of bacterial pathogens in produce. *Postharv. Biol. Technol.* 15:305-311.

Calderon, M. and R. Barkai-Golan, eds. 1990. *Food preservation by modified atmospheres*. Boca Raton, FL: CRC Press, 402 pp.

Calverley, D.J.B. 1996. A *study of loss assessment in eleven projects in Asia concerned with rice*. Rome, FAO.

Campbell-Platt, G. 1987. *Fermented Foods of the World: Dictionary and Guide.* Stonam, Massachussetts: Butterworth Heineman.

FAO. 1986. *Improvement of Post-Harvest Fresh Fruits and Vegetables Handling. Regional Office for Asia and the Pacific*. Maliwan Mansion, Phra Atit Road, Bangkok, 10200, Thailand.

Fellows, P. and Hampton, A 1992. *Small-Scale Food Processing A Guide to Appropriate Equipment.* London: Intermediate Technology Publications. 158 pp.

Food and Agriculture Organization of the United Nations. 1981. *Food loss prevention in perishable crops*. Rome: FAO Agricultural Series Bulletin 43, 72 pp.

Food and Agriculture Organization of the United Nations. 1989. *Prevention of food losses: fruit, vegetable and root crops: A training manual.* Rome: FAO, 157 pp.

Gorny, J.R., ed. 2001. *Food safety guidelines for the fresh-cut produce industry*. Fourth edition. Alexandria, VA: International Fresh-cut Produce Association. 218 pp.

Grierson, W. 1987. *Postharvest Handling Manual, Commercialization of Alternative Crops Project*. Belize Agribusiness Company/USAID/Chemonics International Consulting Division, 2000 M Street, N.W., Suite 200, Washington, D.C. 20036.

Hardenburg, R.E., A.E. Watada, and C.Y Wang. 1986. The commercial storage of fruits, vegetables, and florist and nursery stocks. *USDA Agricultural Handbook* 66.

Harris, K.L., Lindblad, C.J. & Al. 1978. *Post-harvest Grain Loss Assessment Methods*. Washington & Londres.

Harvey, E. et al. 1990. Harvesting and postharvest handling of papayas in the Caribbean. Bridgetown, Barbados: Inter-American Institute for Cooperation on Agriculture (IICA).

Kader, A.A., D. Zagory, and E.L. Kerbel. 1989. Modified atmosphere packaging of fruits and vegetables. *Crit. Rev. Food Sci. Nutr.* 28:1-30.

Kader, AA (ed). 2002. *Postharvest Technology of Horticultural Crops* (3rd Edition). UC Publication 3311. University of California, Division of Agriculture and Natural Resources, Oakland, California 94608. 535 pp.

Kitinoja, L. and A.A. Kader. 2002. *Small-scale post-harvest handling practices: A manual for horticultural crops.* Fourth edition. Davis: University of California, Postharvest Horticulture Series 8E, 260 pp.

Kitinoja, L. and J.R. Gorny. 1999. *Post-harvest technology for small-scale produce marketers: Economic opportunities, quality and food safety.* Davis: University of California, Postharvest Horticulture Series 21.

Knee, M., ed. 2002. *Fruit quality and its biological basis.* Sheffield: Sheffield Academic Press, 320 pp.

Lamikanra, O., ed. 2002. *Fresh-cut fruits and vegetables: science, technology, and market.* Boca Raton, FL: CRC Press, 456 pp.

Lurie, S. 1998. Postharvest heat treatments of horticultural crops. *Hort. Rev.* 22:91-121.

McGregor, B.M. 1989. *Tropical products transport handbook.* USDA Agr. Handb. No. 668, 148 pp.

Mitchell, F.G., Guillon, R., R.A.. Parsons. 1972. *Commercial Cooling of Fruits and Vegetables.* U.C. Extension Manual 43, University of California, Division of Agricultural Sciences, Oakland, California 94608.

Mitra, S., ed. 1997. *Post-harvest physiology and storage of tropical and subtropical fruits.* Wallingford, UK: CAB International, 423 pp.

Moline, HE. 1984. *Postharvest Pathology of Fruits and Vegetables: Postharvest Losses in Perishable Crops.* U.C. Bulletin 1914, University of California, Division of Agriculture and Natural Resources, Oakland, California 94608.

NRI. 1994. *Training Manuel on Grains*. Rome, FAO.

Pantastico, Er. B. 1980. *FAO/UNEP Expert Consultation on Reduction of Food Losses in Perishable Products of Plant Origin.* (6 May 1980, FAO,, Rome). 54 pp.

Paull, R.E. and J.W. Armstrong (eds). 1994. *Insect pests and fresh horticultural products: treatments and responses*. CAB International, Wallingford, UK, 360 pp.

Proctor, F.J. 1985. Post-harvest handling of tropical fruit for export. *The Courier* 92: 83-86.

Rangarajan, A, et al. (no date). *Food Safety Begins on the Farm. A Grower's Guide: Good Agricultural Practices for Fresh Fruits and Vegetables.* Ithaca: Cornell University. 28 pp.

Seymour, G.B., J.E. Taylor, and G.A. Tucker, eds. 1993. *Biochemistry of fruit ripening.* London: Chapman and Hall, 454 pp.

Shewfelt R.L. and Prussia, S.E. (eds). 1993. *Postharvest Handling A Systems Approach.* San Diego: Academic Press, 358 pp.

Smilanick, J.L 1995. *Status of postharvest fungicides and growth regulators.* Perishables Handling Special Issue No. 82: 30-32.

Snowden, A.L. 1990. A color atlas of post-harvest diseases and disorders of fruits and vegetables. Vol. 1. *General introduction and fruits.* Boca Raton, FL: CRC Press, 302 pp.

Sommer, N.F., R.L. Fortlage and D.C. Edwards. 2002. Postharvest Diseases of Selected Commodities. In: Kader, AA (ed). *Postharvest Technology of Horticultural Crops*

Thompson, A.K. 1996. *Post-harvest technology of fruits and vegetables*. Oxford: Blackwell Science, 410 pp.

Walker, D. J. 1992. *World Food Programme Food Storage Manual* Chatham, UK: Natural Resources Institute.

Wills, R., B. McGlasson, D. Graham, and D. Joyce. 1998. *Postharvest: An Introduction to the Physiology and Handling of Fruit, Vegetables, and Ornamentals.* CAB International, Wallingford, UK, 262 pp.

Zagory, D. 1999. Effects of post-processing handling and packaging on microbial populations. *Postharv. Biol. Technol.* 15:313-321.